Steps Towards A Green World

Climate Change and Global Warming

S.K. Bhatia

VIRTUOUS PUBLICATIONS
New Delhi

Revised Edition : 2018

Editor in Chief: Vijay Gupta

Mktg. & Sales :	B-4, Ground Floor, Sector 60, Noida 201301, U.P., INDIA		
Telephone	: 0120 - 4253470, Fax : 0120 - 4253478		
Website	: www.virtuouspublications.com		
E-mail	: contact@virtuouspublications com		

Printed at: VPS Engineering Impex (Pvt) Ltd., Noida Email : vps_8hc@yahoo.co.in

Preface

Without a doubt, Global Warming and Climate Change are the most disturbing news of our times. Alarming consequences are predicted. More than two billion residents of the world, mostly the poor ones in poor countries, will be severely affected. Everyone must learn about the two daunting phenomena of our times; climate change and global warming and their causes, (Greenhouse Effect and Greenhouse Gases). Surprisingly, the living styles of the rich nations are the root cause of the environment's degradation.

This book provides authentic data and reasons thereof, by scientists of world renown whose findings are contained in the assessment reports of IPCC, and UNFCCC, the two world bodies, informing and guiding our needs and actions towards the environment, in which humans, other species and plants can survive in good health. Kyoto Protocol which became the common cause of 196 nations to halt the injurious impact of global warming is dwelled upon in detail (3 chapters) Kyoto Protocol remains in operation, somewhat laconically. Thereafter, the Paris Treaty, signed in December 2015 (still to be ratified by nations) will operate. The features of Paris Treaty, especially affecting India, are detailed in another chapter. But, the principal aim of the book is to dwell upon "measures and actions" which governments, industries, civic bodies, and individuals must perform to save our earth from extinction. The data, descriptions and suggestions for better life styles, is followed by Solar and other forms of renewable energy resources to curtail emission of harmful gases. Technologies, which will change the energy scene are explained. In the final chapter of Part I of the book, the author, gives his personal views and suggestions to make earth's environment friendly and sustainable!

The Part II of the book, is primarily to help students of Environmental Sciences, and Geography. Chapters therein give the genesis of "the climate" formation, since the beginning of this universe. And the various climatic phenomena which befall on our Earth, like hurricanes, rains, sun's radiative energy, ozone zones, Montreal Treaty etc. are described. These chapters are extracted, and kept intact, from authentic sources.

The objectives are: a) the old must know the possible legacy of their life styles, particularly the rich ones, b) The generation, presently in command, must change their living habits as recommended and c) the youth are warned.

Fortunately, the world leaders have realised the consequences of ignoring 'Global Warming'. They have committed their nations, to bring about the changes as prescribed by UNFCCC, at Paris in December 2015. One can believe that all of 196 nations will, continue to improve upon their commitments in the coming years. Hopefully, the life on our planet, will remain environment friendly. And alongside, the developing and under developed countries, will continue with their economic development. Amen!

Technologies and technical terms are explained in simple formats. Brevity, authenticity and readability are the key to the style of this book. Do write to me, about this book's shortcomings and, also, if readers found the effort of mine satisfying!

I take this opportunity to acknowledge and thank the wikipedia, UNFCCC+ IPCC officers in Geneva and London who provided me the IPCC - V reports of WG I,II, & III plus the Assessment Report. I also extracted useful information from newspapers – Indian Express, Hindustan Times and Hindustan (Hindi) and from articles in India Today and The Outlook magazines. My sincere thanks to their contributors and publishers.

I invite readers to suggest corrections and improvements.

S.K.Bhatia
skbhatia1929@yahoo.com

Dedicated

to

(Late) SHRI HIRA NAND BHATIA

My Father

An Example Par Excellence
Who Got Me Educated
By The Best Teachers And Humans

And to

The Youth of India

who will
Lead Us To Economic Prosperity and a Pollution
Free Environment Leaving Behind a Legacy

of

Growth And Sustained Ecology

About the author

Mr. S.K. Bhatia is an alumni of the Indian Institute of Science, Bangalore. For six and a half years, he gained practical experience, in manufacturing electrical equipment in the United Kingdom. Simultaneously, he studied, Advanced Electrical Engineering and Industrial Management, at the Manchester College of Science and Technology, U.K.

He joined the Bharat Heavy Electricals and rose to become its General Manager of Exports. He was loaned to the Ministry of Industry, Government of India, to become Director General of, The Research and Development Organisation of Electrical Industry. Later, in the private industry, he served as a Senior Vice-President (Technical) in Avery India Ltd., a subsidiary of the General Electric Corporation, England. He also became a Board Director of its another subsidiary, Schenck-Avery Ltd. specialising in the manufacture of 'testing equipment' for the Auto Industry.

Mr. Bhatia has written a book on management, "People Who Make Profits". His book "भूकंप के झटके" won the prize of the Hindi Academy of Delhi. This book, has been translated in to English, and Marathi. Many of his stories, both English and Hindi, won prizes in all-India competitions.

Contents

Part 1

Part II

PART I

Climate Change and Global Warming

Includes

Polluting Life Styles
Role of World Bodies (UNFCCC & IPCC)
Latest Technologies, Remedial Measures
Economic Growth and Sustained Ecology

1

CLIMATE CHANGE AND GLOBAL WARMING – THE ISSUES

What is Climate Change?

A read of an anonymous, possibly a Scandinavian sailor's diary, on fishing expeditions in the Arctic seas, may be helpful.

"1979. My father carried me on his boat to fish in the Arctic ice lands. I was just 13 years old. Occasionally we spotted ' reindeers pulling sledges carrying families of Eskimos and dogs running by their sides. There was solid ice for miles beyond our eyes could see . Off and on ' Polar' bears appeared in their brown coats, catching fishes in broken ice-ponds. Penguins too were frolicking.

"1991. I went to those lands of ice on my own. Neither reindeers nor Eskimos or polar bears came my way. Ice lands were broken here and there and Penguins were fishing. Broken icebergs floated at many locations. I had to be vigilant to navigate and avoid crash-landing.

"2010.My father is no more. But I still ferry my boat in the Arctic lands. There are floating ice-bergs all around. Though fish are in plenty, I miss my old pals; reindeers, dogs, polar bears, even penguins."

Perhaps climate is changing.   *"Anonymous"*

It is a telling story of the rising temperatures over the previous forty years. Ice bergs were melting. Animals and aquatic species were vanishing or possibly moving to remoter parts.

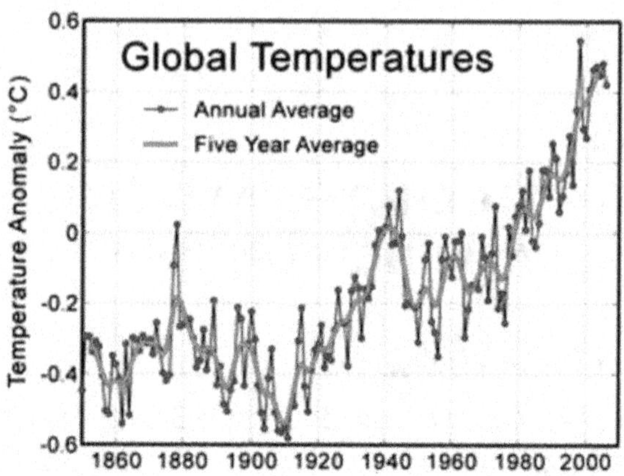

Fig.1.1, The instrument recorded temperatures from 1850
[coloured version of the fig is on page 233]

Weather Experts Views and Predictions

Weather experts believe that Global Warming is happening since 1950s. The heat absorbed within the earth till 1970 oozed out and set the temperatures rising; in the atmosphere and ocean waters Fig.1.1, supports the rising temperature opinion.

Predictions: Climate Change means rising temperature, greater frequency and higher magnitude of rains, hurricanes and cyclones. The predictions, are "very likely", 95% IPCC experts back these. However, the years of happenings, their intensity or numbers may vary.

Researchers claim that devastating rain-cum wind upheavals e.g. "Grey Swans" may be anticipated by combining physical knowledge with historical data. "Grey Swan" tropical cyclones can whip up catastrophic surges bigger than what can be predicted through weather data alone. However, they can be forecast if data is supplemented by global simulations. Researchers have recently found, through simulations, that they could strike "Dubai", where never before a tropical storm ever has been recorded. In Tampa (Florida, USA) and Cairns Australia, the frequent storms were likely to rise to unprecedented magnitudes in the next century. The world has not seen a 'black swan' or 'grey swan' cyclone. ("Nature, Climate Change- 31 August 2015. Authors Ning Lin and Kerry Emanuel)

Greg Holland, of US National Climate Research Department, divides ocean storms in 3 periods:-

(a) 1905-1930. Annually 6 ocean storms; of which, 4 of hurricanes (wind speed> 118kms/hour)

(b) 1931-1994. Annually 10 ocean storms, of which, 5 of hurricane intensity.

(c) 1995-2005. Annually 15 ocean storms, of which, 8 of hurricane intensity; including hurricane "Katrina" which devastated Louisiana.

North America can expect more hurricanes, floods, droughts, heat waves and wildfires, the report said, and the coasts will be flooded by rising sea levels. Crop production will increase initially as the growing season gets longer, but climbing temperatures and water shortages will ultimately lead to sharp reductions of foods and fodders. Food production will fall by half in many countries and governments will have to spend 10% of their budgets or more to adapt to climate changes, the report, of IPCC, said.

Asia will suffer from unprecedented flooding as the rising temperatures melt Himalayan glaciers and rock avalanches will wipe out many villages. The same will happen in the European Alps and the South American Andes.

Rising temperatures and drying soil will replace the moist rain forest of the eastern Amazon with drier savannah, eliminating much of the habitat that now supports the greatest diversity of species in the world.

Drowning Islands and Cities in Seas

The Republic of Kiribati, is a nation state consisting of 33 atoll and reef islands in the Pacific Ocean, across the 'date line'. Dispersed over 3.5 million square kilometers, they straddle the equator. A just over 100,000 (2011) people live on them, half of whom live on Tarawa Atoll. These islands are likely to go under rising sea waters in this century.Ten more areas likely to "sink" under waters are (i) Maldive Islands, Arabic Sea, (ii) Bangkok, Thailand, (iii) Mississippi River Delta, USA, (iv) Komooto Island, Indonesia, (v) Great Barrier Reefs, Australia, (vi) Tokyo, Japan, (vii) Venice, Italy, (viii) Thames, London, U.K. (ix) Big Sur, California, USA (x) Panama Canal, Panama. Later, the drowning of seven villages in Odhisa, India, has been mentioned.

Antarctica, Arctic

Glaciers in western Antarctica are moving five times faster than before and are discharging water into the ocean around them. Combined with Arctic ice-melt discharge, the global sea-level rise is now 3mm per annum, against the decade ago rate of rise of 2 mm in a year. The Ilulissat Glacier, which presumably sank the Titatanic, " ... has seen a massive acceleration of the speed ....the ice is moving at 2 meters per hour on a front of 5km long x 1500 meters deep.... One glacier puts enough fresh water into sea in one year to provide drinking water for a city of the size of London". The sea level of Europe's coastline could go up by 2 meters; the IPCC estimate of 20-60 cm rise every decade is based on two years old data. Dr. Corell, Chairman of the Arctic Climate Impact Assessment, while flying over the 'Llulissat' glacier saw 'gigantic holes' in it through which swirling masses of melted water was flowing. These melt-water rivers are lubricating the glacier, like applying oil to a surface. The massive acceleration could be catastrophic." The glacier is now moving at 15 km a year into sea.

Veli Kallio, a Finnish scientist, said that the quakes, of 1 to 3 magnitude, were triggered because ice had broken away after being fused to the rocks for hundreds of years.... The quakes though not dangerous show that ....events are happening far faster than ever anticipated."

Tensions are growing among, primarily, USA, Canada, Russia and Denmark on account of the melting, during summer, of the seas in the Arctic Ocean. As a consequence a North- Western waterway opens for navigation avoiding longer distances to the Panama canal, for the container traffic between the Atlantic and Pacific ocean countries. Moreover, there are huge mineral and oil reserves beneath those waters, for which these countries are placing their claims.

Great Barrier Reefs, "Worst mass bleaching ever recorded".

The Great Barrier Reefs lie along the coast of Australia from Cairns to the island of Papua New Guinea. Their multicoloured corals were a delight to watch. A recent aerial survey of more than 500 coral reefs show an appalling sight; they are "Bleached". Professor Terry Hughes, convener of the National Coral Bleaching Taskforce states, "We flew over 4000 km... and saw only four reefs that had no bleaching. The severity is much more than .... in 2002 or 1998..... The fact that these hardy species have turned white shows just how severe summer conditions have become on the northern CBR. Says Hughes, "Scientists .... reporting 50% mortality of the bleached corals..... but too early to tell just what the overall outcome will be".

Coral bleaching occurs when abnormal environmental conditions like heightened sea temperatures cause corals to expel tiny photosynthetic algae called "zooxanthellae". Bleached corals can recover if the temperature drops and zooxanthellae are able to recolonise, otherwise the coral may die.

(Reported by PTI, Melbourne March 29, 2016)

Summary of IPCC Assessment Reports IV (2007) And V (2014)

IPCC report in January 2007, forecast that average temperatures are likely to rise by 7-10 °C in half a century. The report characterised global warming as a runaway train that is irreversible but its speed can be moderated by societal changes. The report said, with more than 90% confidence that, the warming is caused by humans. Its conclusions were based on, and widely accepted, the years of accumulated scientific data supporting it.

A more detailed report on global warming was issued by the United Nations, in April 2007. It painted a near-apocalyptic vision of the Earth's future if temperatures continue to rise unabated. Arctic and Antarctic glaciers will melt leading to rise in the level of oceans and seas. Low lying coastal areas will be lost forever; some happenings and those forecasted are listed above. Ice on mountains and glaciers will turn into water, will flood rivers for a season or two. Thereafter, rivers will dry and hardly sufficient water will remain for irrigation or drinking. The report foresees more than a billion people in desperate need of water, extreme food shortages in Africa and elsewhere, a blighted landscape ravaged by forest fires and floods, at least 30% of the world's species will disappear if temperatures rise 3.6 degrees above the average levels of the 1980s and 1990s!

Representatives of some of the world's largest Greenhouse Gas, GHG, emitters attempted to tone down the report. But, scientists fought for their predictions and had them included in the IPCC reports. United Nation's Committees of scientists and weather experts conclude that the phenomenon of global warming, because of human living styles, is rampant. Rich Nations are concerned; (a) underdeveloped and developing countries may demand hefty financial compensations, on account of the industrialised nations' emitting high amounts of GHG and (b) reducing emission of GHGs will/may curtail the life styles of their populations.

However, the IPCC Assessment Report V 2014, (synopsis of reports given in a later chapter), predicts 3 to 6 °C temperature rise by the year 2100. Though, the future remains grim as predicted in Report IV, the alacrity of earlier Report gets heightened, in spite of the mitigation and adaptation measures taken by many countries in the past 20 odd years.

A recent (October 2015) report by the UN climate body predicts higher temperature rise than the targeted 2 °C temperature rise, by UNFCCC. This view comes after experts had scrutinised INDC plans submitted by a large majority, 146-170, of countries. The report indicates that by 2100, the rise of temperature would be 2.7 °C. In the absence of these INDC plans the rise in temperature by then could be between 3 °C to 6 °C. The report indicates that emissions would decline by 9% by 2030, if the INDCs get implemented. (Amitabh Sinha, Indian Express 30-10-2015 quotes Christiana Figures of UNFCCC).

Some assessments believe that winters are getting warmer by 1.1 °C to 1.6 °C annually.

Impacts of Rising Temperature

1. The increase of temperatures is conducive to a rise of epidemics and doesn't facilitate sanitary conditions; e.g. epidemics such as cholera, malaria, dengue fever will increase in some areas. In the Andes Mountain in Columbia for example, mosquitoes that can carry dengue and yellow fever viruses were previously limited to 3,300 feet (1,006 m) but recently appeared at heights of 7,200 feet (2,195 m).

2. Environmental migrations will be the source of a new category of refugees or internal displaced populations;

(IDPs), voluntarily or forced by the governments. Nearly 10 million people were displaced in year 2006, and will create up to 50 million environmental refugees by the end of the decade.

The Red Cross says, 'environmental disasters have already displaced more people than wars'. Environment - related migration has been most acute in sub-Saharan Africa, but also affects millions of people in Asia and India. Europe and US face increased pressure from people driven from North Africa and Latin America by deteriorating soil and water conditions. New Zealand has already agreed to accept the 11,600 inhabitants of the low-lying Pacific island state. Elsewhere, as many as 100 million people live in areas that are below sea level or liable to storm surge. A total of 213 communities in Alaska are threatened by tides that creep three meters further inland each year. The climate experts, by consensus, predict doom on the planet by 2050, in case mitigation efforts get thwarted.

3. The scarcity of resources could exacerbate tensions already existing between different ethnic groups. There can even be a war, especially for securing larger resources of drinking water.

4. Mr. Al Gore former Vice-President of the USA foresees disastrous future and has been urging nations to stall the unpleasant consequences of Climate Changes by judicious correction of their life styles; because humans are the principal cause of environmental pollutions. He spearheaded a campaign, on global level, to halt emissions of GHGs.

Mr. Gore produced a film "An Inconvenient Truth", based on his book of the same name, warning this world of the harmful impacts of global warming. He is the founder, and current chairman, of Alliance for Climate Change.

Mr. Al Gore and the IPCC (headed by India's Mr. R.K. Pachauri) were awarded, jointly, the Nobel Peace Prize for year 2007 for their efforts in highlighting the impacts of Global Warming cum Climate Change.

Mr. Blair the former Prime Minister of the UK called upon his countrymen to heed to the warning of global warming. The British Government, set in measures to restraining emission of greenhouse gases to mitigate the ill omens which may visit our children and grandchildren. The majority of countries subscribe to the United Nations consensus to reduce the greenhouse gases to save the

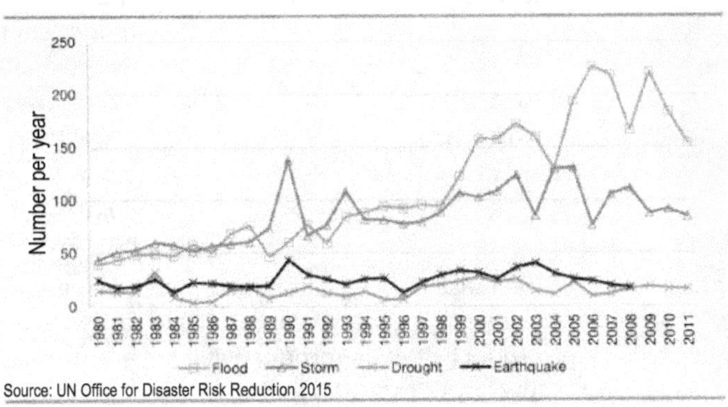

Fig.1.2, The frequency of floods has increased three-to four-fold
[coloured version of the fig is on page 233]

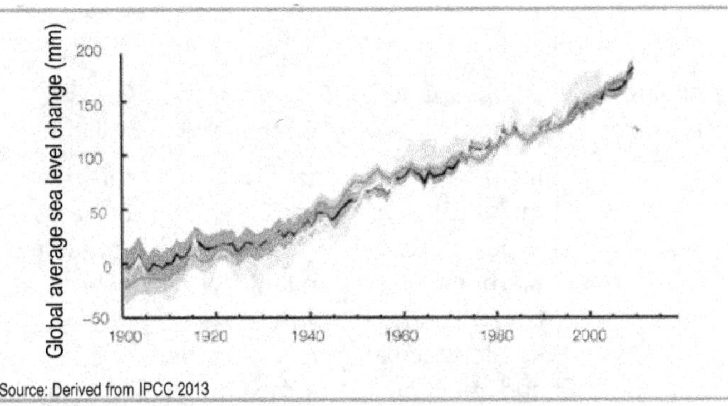

Fig.1.3, Sea levels has been rising since the late 1800s
[coloured version of the fig is on page 234]

world, but few serious attempts are on hand to halt the degrading situation.

A report of IPCC predicted, in the table below, results as the global temperature rises.

Table 1.1 Impact of Temperature Rise Values vs. Likely Impacts on the Earth

Temperature Rise	0 °C to 1 °C	1 °C to 2 °C	3 °C to 4 °C & 3 °C to 5 °C
WATER SUPPLY	Increase in High latitudes and wet tropics. People Exposed to increased water supply	Decrease in mid latitudes and dry tropics likely Population affected 2 bn	Decline of water stored in glaciers and snow cover will reduce supply where one sixth of world population lives. Population affected 3bn
FOOD	Decrease in cereal productivity at low latitudes >	Some increase in cereal productivity in mid to high latitudes due to more rainfall and warmer weather. >	Decrease in all latitudes >
Extra People at Risk of Hunger	0 °C < 2 °C 50 million	2 °C & >3 °C 132 million	3 °C < 5 °C 266 million
HEALTH	Increased malnutrition, exposure to diarhoea, cholera and cardio-vascular and infectious diseases	Increased mortality, diseases and injury from heat waves, floods, storms, fires and droughts	Infectious diseases spreading to new areas

ECOSYSTEMS Ditto	Up to 30% plant and animal species at increased risk of extinction. Increased Coral bleaching	Higher percentage at Risk Most Corals bleached at < 2 °C	Over 40 % of species become extinct
COASTS	Increased damage from floods and storms	Millions of more people will experience more coastal flooding each year	Around 30% of coastal wetlands lost

While survival and well-being of humans depends heavily on foods; vegetables, fruits, meats etc, on the other hand, the bacteria in their digestive tracts also affect their lives. Small quantities of non-living nutrients, such as phosphorus, iron and calcium help to maintain blood circulation and digestion, but higher concentration of any of these can be harmful. Interactions among humans, animals, and plants also have a bearing on the environment and climate.

Those Affected

The devastating effects will hit populations of poor nations, mostly in south Asia and Africa. Millions are without the resources to meet the challenges of climate changes and vagaries of weather. Such countries lack technologies and finances to adopt, and to adapt to the adversities of forthcoming changes. Industrialised countries must come forth with programmes of technologies' transfer and liberal back up of funds.

Skeptical scientists (about 5%) say that it has not yet been ascertained whether humans are the primary cause of global warming. They attribute global warming to natural variation of ocean currents, solar activity, cosmic rays or unknown natural causes.

Computer Models: Computer models employed to predict climate changes are flawed; even the fastest super-computers cannot simulate the formation of clouds and other climate related complexities.

i) Solar Activity: Solar activity is more significant in global warming than human caused emissions of greenhouse gases. Increase in solar activity, translate into fewer cosmic rays, fewer clouds and more global warming.

ii) Projected Temperature: Increases do not translate into an extreme threat to humanity because they just mean moving from San Jose to Los Angeles, which poses no significant environmental danger.

iii) Evidence: From ancient records of Nile floods to modern analysis of polar ice cores shows that most of the Earth's recent warming occurred before 1940, before emissions of GHG caused by human activities.

End of Controversy

The controversy, about causes of Climate Change between (i) proponents of human activities which produce greenhouse gases and their detractors who (ii) believe that the global warming is due to sun's varying radiated energy, is likely to end with the former ruling out conclusively the latter, as per researches of Mike Lockwood, of the Rutherford Appleton Laboratory, Oxford Shire UK. He advocates, "The temperature record is simply consistent with any of the solar- forcing that people are talking about. The data collected in the last 25 years of the 20th century has put the notion to rest. Though the sun's activity has been decreasing since 1985, global temperatures have continued to rise at an accelerated pace, instead of slowing down. The top 10 warmest years have happened in the last 12 years. Since 1900, the mean global atmospheric temperature has risen by 0.8 °C and the sea level by 10-20 cm. While the scientist circles are agreed about the role of human activities as the main driver of climate change, a recent poll suggested that the public still believes, there is significant scientific uncertainty. (Hindustan Times July 12, 2007)

Global Warming, Cause of Climate Change

Climate experts, backed by researches, assert that the principal cause of climate changes is the world-wide 'Global Warming'. Their findings further warn that human life styles, (primarily high consumption of energy in Industrialised countries), plus ambitious plans of industrialising of China and India, will warm up our world appreciably. While the processes leading to Global Warming are dealt with in detail in the following chapters, the evidences to support rising temperatures have been illustrated above.

The Issues to Ponder

Accepting that the planet is getting warmer, the common man must know:

1. What causes Global Warming ?

2. Who are the principal contributors to Global Warming ?

3. What efforts have been done by 'World Bodies' to halt Global Warming?

4. Measures to better life styles and Eco-Sustainable Economic Development

Ultimately, one must heed to the "Wake-up Call" by UNFCCC and IPCC against the likely damaging consequences of Global Warming. The poor, though deprived of facilities of modern era, at least must have a fair share of God given sources of food and healthy environment to live decently till they survive. Above all, we-the present generation-must leave behind a healthier planet for our children and grand children.

Appendix 1.1

Significant Climate Anomalies and Events in 2015

i)   **Heat Events**: Canada, the warmest summer, South America, the warmest, since records maintained, Asia the warmest since 1910. USA and Europe second warmest year since 1895 and records kept respectively. Alaska, second warmest year since 1925.

ii) **Rainfall Events**: USA, wettest year since 1895, and wettest May ever. Mexico, wettest March since records maintained, 1941. Marrakesh, Morocco, received on 6th. August, 13 times the average, annual, rainfall in just one hour. India, Chennai had the wettest December, after 100 years. Nearly 200 people died. Floods in China affected 75 million people. Southern China had the wettest May in 40 years.

iii) **Wind Events**: Three hurricanes, (Kilo, Ignacio, Jimena) occurred simultaneously in Eastern North Pacific Ocean since records maitained-1949. Off Mexico, the strongest tropical storm in the Western hemisphere, hurricane Sandra in November.

iv) **Ice events**: Arctic saw the smallest annual maximum extent of sea ice. In the melt season, there was the fourth smallest minimum extent. The Antarctic during the melt season, reached the fourth largest minimum extent of sea ice on record.

Source: US Department of Commerce, National Oceanic and Atmospheric Administration.

va) "Indian Express" an English daily of Delhi, reported in its 27-05-2015 edition "A Heat Wave in India" and a report followed that more than 400 persons died due to heat waves. in different parts of the country.

vb) On 28-05-2015 a Hindi daily "The Hindustan" carried a headline, " 2015 to be the hottest ever year of the universe". During the year the temperature to rise by 0.80 °C as against the predicted, by 'meteorologists', a rise of 0.64 °C. The paper attributed the cause of the temperature rise to 'GLOBAL WARMING' because the emission of CO_2, above 400 ppm was recorded, for the first time on a day in 2015 (Excerpt translated).

vc) Vidharbha in Maharashtra experienced draught for the third time in a row, resulting in the highest number of suicides by farmers.

vd) In December 2015, rains descended on Chennai, in torrents for days on end. Humans died in hundreds and properties were decimated on large scale; an experience after 100 years as per records in the Met. department.

vi) USA Storms hit central US states. At least 43 persons got killed and hundreds of buildings were flattened. Generally, such weather happens in those parts in spring and summer. Extreme weather conditions happened across the country: heavy snow in Mexico, Texas and Oklahoma and flash floods in parts of the plains and mid-west. However, the East Coast experienced an unusually warm weather; warmest Christmas Day in history in New York; 66 °F and 72 °F.

vii) South America experienced the 'Worst' floods in 50 years in Paraguay, Argentina, Uruguay and Brazil. El Nino phenomenon caused floods in three major rivers displacing nearly 150,000 persons from their homes. Paraguay declared a state of emergency; Argentina created a 'crisis committee'.

viii) U.K. rivers overflowed their banks, due to incessant rains, in Yorkshire, Lancashire and Greater Manchester. The severity of water flows was so high that the UK's Environment Agency recommended a "complete rethink" of the UK "Floods Defence Measures".

ix) On the other hand the Alpine resorts in Austria, France and Switzerland experienced warmer climate than usual. Skiing enthusiasts had to forgo their New Year holiday enjoyment due to "the balmy weather"; it happened the same last year, too. The 'city of canals'- Venice Italy- experienced unusually low tide. Usually submerged parts came up surface on Sunday (Christmas).

x) Philippines suffered from typhoons more than once. Losses of lives and property were colossal.

xi) Portugal was devastated by floods like never before.

xii) The vagaries of climate are on the increase. Rising temperatures will endanger life in the Gulf region, say scientists. By the year 2100, some centers in the Middle East are" likely to experience temperature levels that are intolerable to humans". The waters in sea will create humidity, which combined with high temperature will create conditions unbearable for humans; sweating to absorb heat will not be possible due to high humidity.

("Nature Climate Change" October 26, 2015 by Jeremy S Pal & Elfatin A B Eltahir).

Happenings Indicating Climate Change Over Long Periods

Every country, region and area, has experienced unpredicted and unprecedented weather conditions, in the recent past. Some examples are given below.

Delhi

Temperatures hover about 40°C in April; the probability of smog increases 100%. Delhi will face a huge water shortage because the Himalayan glaciers are depleting at a faster rate. Residents of Dwarka and many other colonies have to wait for another year or more to get adequate potable water supply.

The poor state of the 'Ridge' and depleting forest cover are a huge cause for concern. The Aravali Ridge is no longer the coolant for Delhi's temperature as it once was; a cushion to Delhi from warm air flowing from Rajasthan. The green region of NCR had depleted by 50% between 1977 and 1997. The proliferation of 'vilayti kikar' has destroyed other vegetation on the Ridge.

Besides depleting green cover, the high rate of emission from 50 lakh plus vehicles are pushing Delhi closer to its doom. Though a speculation so far, 2015 year's heat wave in the plains of Northern India, are due to global warming, according to weather scientists.

And yet in January 2015, Delhi's citizens woke up to frost covered streets. It was the first after 70 years. The white was mesmeric for warm-bedded ones but unkind for road dwellers. Visibility being low, air passengers simply cursed the Gods!

India

The instances of unexpected and oddities of weather are happening at many places in India. July 26, 2005, Mumbai, recorded about 970 millimeters rainfall, the highest ever downpour in 24 hours. The city was submerged under water, low-lying areas and cars were drowned. The office workers were locked up in their work-places for 48 hours, unable to commute to their homes. Schools were closed. Nearly one thousand persons were killed and property worth 4,000 crores rupees was damaged. A repeat, virtually, took place the following year.

In 2006, the roads and colonies in Bangalore and Chennai suffered no less from unprecedented rains. But Rajasthan and parts of Maharashtra had remained dry for 2 to 3 years causing misery to millions of farmers. Can we forget the devastation by Tsunami waves in 2004, never experienced before in India?

Hyderabad's average rain in May is 40mm. May 2003 passed away without a drop fall. In June the temperatures soared to 47 °C. One Lakh Muslims' congregation prayed for Allah's blessings. 5000 Hindus chanted to their Lords for water. Only then it started to rain, but 1300 people had died till then.

Shimla received 108.5mm snowfall in February, 2007. Pine and Deodar trees covered under thick white blankets, were heavenly for honeymooners. But less adventurous tourists were stranded, mobile networks were stalled and power lines broke down. It was hell, though a century's marvel!

2013 experienced unprecedented rains, floods in Uttarakhand. The religious shrine of Kedarnath was reduced to ruins. 2014 saw a similar fury of rains and floods in Srinagar. And 2015 November went through catastrophic rain fall for days on. Chennai was submerged under water. In all the three calamities, the resulting devastation ruined houses and residents had to go through unbearable miseries which have never been experienced in generations.

Orissa lost, to Bay of Bengal waters, a few villages. Gobindpur, Manhipur, Kuainri Ora villages went under sea in 1980s. In 1990s, the sea devoured two villages, Kharikula and Sarpada. The sea water has come inside the land by 2.5 kilometers. Satbhaya (seven villages) and Kanhupur areas are leading precarious existence. More villages are awaiting extinction.

'Besides global warming, the loss of mangroves, near Mumbai, Kolkata and Chennai, has led to loss of drainage causing floods in those cities. Land is eroding faster.

Director Nila Madhab Panda of the Discovery Channel telecast the film, in 2006. " Climate's First Orphans". (Report in HT of 18-5-2007). It is mind shattering.

USA & Europe

The USA and Europe are no less affected by changes in climate in their region. Year 2005, New Orleans, USA was hit by hurricane 'Katrina'. Winds of 250 kilometers per hour ravaged the city. More than a million refugees left New Orleans because their houses collapsed and cars were drowned under high waters. Hurricane Patricia in Mexico, in October 2015, surpassed wind speeds, 320 kilometers per hour. And recently, Texas faced a deluge due to heavy, unprecedented, rains. In living memory of older people, the frequency of tornados, hurricanes and typhoons in Florida and nearby seas has increased.

The USA and Europe have also suffered heat waves during summer and colder winters. For example, in the year 2006, Europe baked under 40 °C. The hottest summer ever, was followed by colder than normal winter in Northern America. There were floods in Scotland and Germany. Spain and Portugal reeled under body deep floods due to incessant rains, in October 2015.

2

GREENHOUSE EFFECT, GREENHOUSE GASES AND GLOBAL WARMING

The "Greenhouse effect", GE,. All of us are familiar with glass houses (or 'green houses') in nurseries. The plants grow under glass/ plastic enclosures/covers to save them from direct sun-rays' heat. Inside those enclosures we experience strong smells and moisture hangs in the aerosol.

In effect, only about 70% of infrared waves, of the sun's radiations, fall on the Earth's surface, which then bounce back into the atmosphere. They mix with 30% infrared waves, already trapped there. The sensible and the latent heat from ground meets moisture in the atmosphere, and water droplets in clouds. Like in greenhouses in nurseries, the described phenomena result primarily in the release of water vapour, carbon dioxide, methane and nitrous oxide. These are called Green House Gases, GHG. Besides the GHG, sulfur hexafluoride and Cloroflorocarbons (CFCs) are also released in minute quantities.

These greenhouse gases, floating in the atmosphere also emit long wave radiations both upward to space and downward to the surface. This phenomenon is called "Greenhouse Effect-GE", which essentially is responsible for rise in temperature, i.e. "Global Warming" on our planet, (as per IPCC). The phenomenon was essential for life on Earth in the earliest times. Without it,the average temperature at the equator would be -10 °C. Fig. 2.1 shows the GE, picturesquely.

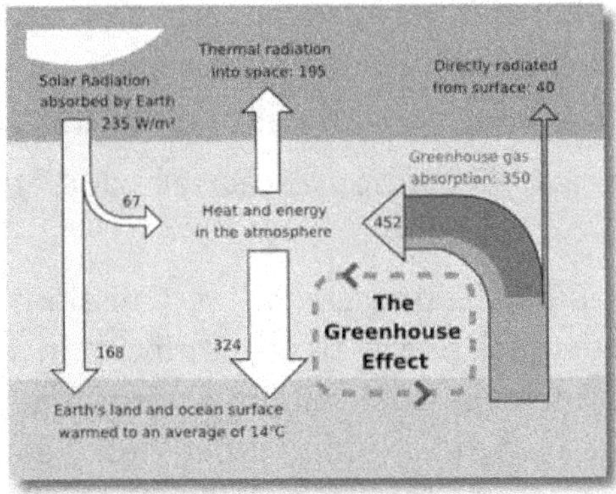

Fig. 2.1, Greenhouse effect-GE

Greenhouse effect was discovered by Joseph Fourier in 1829 and quantitatively measured by Svante Arrhenius in 1896.

Green House Gases and Other Constituents in the Atmosphere

When ranked by their direct contribution to the Greenhouse Effect, the most important are:

Compound	Formula	Contribution (%)
Water vapour and clouds	H_2O	36–72%
Carbon dioxide	CO_2	9–26%
Methane	CH_4	4–9%
Ozone	O_3	3–7%

Non-Greenhouse Gases e.g. Carbon monoxide (CO), or Hydrogen Chloride (HCl), besides Nitrogen (N_2), Oxygen (O_2), Argon (Ar) etc. are present in the atmosphere. Their contribution to GE, or warming is insignificant.

Other than the atmosphere, oceans too absorb huge amounts of carbon dioxide. Warmed up sea waters do contribute to GE, as described later in this chapter.

Contribution of Clouds to Earth's Greenhouse Effect

Clouds are major non-gas contributors to Earth's GE. Clouds contain water droplets or ice crystals suspended in the atmosphere. They also absorb and emit infrared radiations and thus contribute to warming up effect, like the radioactive properties of the greenhouse gases.

Accumulation of Greenhouse Gases due to Human Activities

Anthropogenic GHG emissions (i.e. emissions produced by human activities) come from combustion of carbon-based fuels, principally coal, oil, and natural gas. Modern transportations, agricultural processes along with deforestation also contribute to release of GHG; which consist mostly of CO_2 and Methane (CH_4) (54.7% and 30 % respectively).

Since the beginning of the Industrial Revolution (taken as the year 1750), human activities have produced 40% increase in the atmospheric concentration of CO_2 from 280 ppm in 1750 to 400 ppm in 2015. This increase happened despite the uptake of a large portion of the emissions by various natural "sinks" (forests absorb CO_2) involved in the

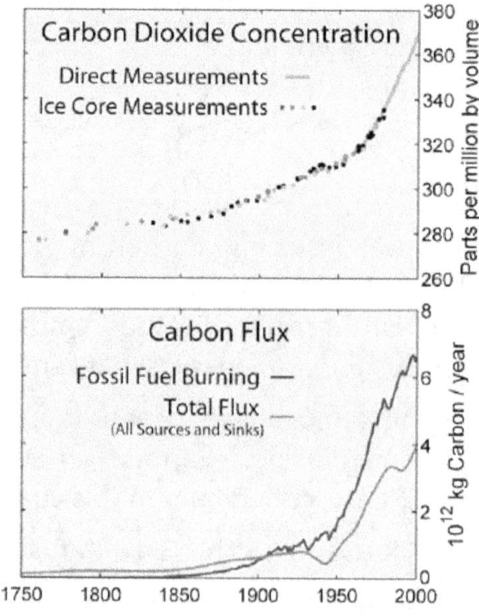

Fig. 2.2, Carbon dioxide and carbon flux

photosynthesis cycle, It is worthwhile to know that, without greenhouse gases, the average temperature of Earth's surface would be colder than the present day average of 14 °C (57 °F). In the Solar System, the atmospheres of Venus, Mars and Titan also contain gases because of the greenhouse effect. The amount of CO_2 releases are graphically shown in Fig. 2.2.

Warming by Greenhouse Gases (GHG)

The GHGs accumulated in the atmosphere, clouds and sea waters, release their absorbed heat, according to characteristics of each gas. Primarily CO_2 and methane have "life"(methane has short life and CO_2 has extremely lasting life). The heat thus given out, raises temperature all over the globe. Water vapour exists in gaseous form in the atmosphere, including clouds. The heat absorbed by water vapour, as a GHG, also adds to increase temperature globally. Warmed up sea waters also absorb CO_2, which in the Arctic and Antarctic sea lands heat up ice-lands. Ice in those lands loosening, gradually turning into waters. These changes in turn are affecting species and life styles of populations living there .

Carbon dioxide reaches its maximum warming effect ten years after its release. And this effect stays in the atmosphere for 1000 years thereafter. The graph in the fig. 2.2 indicates increasing atmospheric CO_2 levels as measured in the atmosphere and ice cores. In the lower picture, the lower curve shows the amount of net carbon increase in the atmosphere, and the other curve shows carbon emissions from burning fossil fuels. Sector-wise contribution of greenhouse gases in India is indicted in the table below.

Note: Two scales can be used to describe the heating effect of different gases in the atmosphere. The first, the atmospheric lifetime, describes how long it takes to restore the system to equilibrium following disappearance of the particular gas. e.g.

interchange with other reservoirs such as soil, oceans, and biological systems. The "mean lifetime" refers to the decaying away of the excess. The mean life time of CO_2 is thousands of years, that of methane is nearer to a hundred years..

Atmospheric concentrations of greenhouse gases are determined by the balance between sources (emissions of the gas from human activities and natural systems) and "Sinks" (the removal of the gas from the atmosphere by conversion to a different chemical compound). The proportion of an emission remaining in the atmosphere after a specified time is the "airborne fraction" (AF). More precisely, the annual AF is the ratio of the atmospheric increase in a given year to that year's total emissions. For CO_2 the AF over the last 50 years (1956–2006) has been increasing at approximately 0.25% per year.

In the early 20th century it was realised that the known major greenhouse gases in the atmosphere did cause the earth's temperature to be higher than it would have been without them; the Earth's average surface temperature is about 20-30 °C warmer than it would be without the Greenhouse Effect (IPCC).

It has been estimated that if GHG emissions continue at the present rate, Earth's surface temperature increase could exceed 3 °C, as early as 2047, with potentially catastrophic, effects on world populations, ecosystems, biodiversity and the livelihoods of people worldwide.

Role of Carbon in Climate Change

Around 1900 Giga tons of carbon are present in the biosphere. Carbon is an essential part of life on Earth. It plays an important role in the human and animal structures, biochemistry, and nutrition of all living cells. Carbon exists in the Earth's atmosphere primarily as carbon dioxide (CO_2), a gas. Although it is approximately 0.04% (on molecular basis) of the atmosphere, it plays an important role in supporting life. Methane (CH_4) and chlorofluorocarbons (CFCs) (the latter is entirely due to humans) also contain carbon. The CO_2 and methane form major part of greenhouse gases in the troposphere. Their increasing concentration, in recent decades, is believed (by a majority of scientists-IPCC) to be contributing to global warming.

As the sea waters warm up, their absorption capacity declines. Melting ice on the other hand results in collected-pools of water, which in turn absorbs more energy. Partly, for this reason, the Arctic is warming faster than other places. The seas in the north of Canada, becomes waterways in summer and are suitable for navigation. Reindeer, Brown Wolf and many aquatic species are going instinct.

Ozone as GHG. An isotope of oxygen, Ozone with 3 atoms of oxygen, is formed by a photochemical reaction, with other gases e.g. nitrous (N_2O) oxide with hydrocarbons (released prominently by auto-vehicles) which requires high intensity light and heat. Over millions of years, small quantities of ozone collected, and formed a layer, called "Ozone Layer" at the lower end of Stratosphere, between 10-15 kilometers above the surface. The ozone shield is important because it protects plant and animal life on land from the sun's ultraviolet rays, which can cause skin cancer, cataracts, and damage to the immune system. Ozone layer also helps preserve the DNA of plants and animals.

The Ozone layer is getting depleted since 1980s and even earlier, thus increasing the possibility of harmful UV rays getting through to earth's surface, causing higher temperatures and risks of adverse health. Montreal Protocol stopped such depletion; see next section. {more details in chapter 4 (Part I) and chapter 7 (Part II)}

CFCs and HFCs in the Atmosphere

Besides Green House Gases, CFCs are also released in the atmosphere, though in miniscule quantities. Acting with other GHGs, CFCs deplete ozone and are thus harmful.

But Du Pont Corporation introduced CFC for refrigeration; their use got extended to air conditioning. By 1976, manufacturers in the United States were producing 750 million pounds of CFCs a year, and finding all sorts of creative uses for them, from propellants in aerosol sprays, to solvents used to clean silicon chips, to automobile air conditioning, and as blowing agents for polystyrene cups, egg cartons, and containers for fast food. "They were amazingly useful," wrote Anita Gordon and David Suzuki. "Cheap to manufacture, non-toxic, non-inflammable, and chemically stable." By the time scientists discovered, during the 1980s, that CFCs were thinning the ozone layer over the Antarctic, they found themselves taking on a $28-billion-a-year industry.

By the time they were banned internationally during the 1980s, CFCs had been used in roughly 90 million car and truck air conditioners, 100 million refrigerators, 30 million freezers, and 45 million air conditioners in homes and other buildings. Because CFCs remain in the stratosphere for up to 100 years, they will deplete ozone long after industrial production of the chemicals ceases.

These human-created chemicals do more than destroying stratospheric ozone. They also act as greenhouse gases, with several thousand times, the per-molecule greenhouse potential of carbon dioxide. What's more, the warming of the near-surface atmosphere (the lower troposphere) seems to be related to the cooling of the stratosphere, which accelerates depletion of ozone at that level. An increasing level of carbon dioxide near the Earth's surface "acts as a blanket," said NASA research scientist Katja Drdla. "It is trapping the heat. If the heat stays near the surface, it is not getting up to these higher levels." (Borenstein)

During the middle 1990s, scientists were beginning to model a relationship between global warming and ozone depletion. A team led by Drew Shindell at the Goddard

Institute for Space Studies, created the first atmospheric simulation to include ozone chemistry. The team found that the greenhouse effect was responsible not only for heating the lower atmosphere, but also for cooling the upper atmosphere. The cooling poses problems for ozone molecules, which are most unstable at low temperatures. Based on the team's model, the buildup of greenhouse gases could chill the high atmosphere near the poles by as much as 8 °C. to 10 °C. The model predicted that maximum ozone loss would occur between the years 2010 and 2019. (Shindell, et al.)

GHG Contributions By Human Activities (Statistical Data)

Global GHG emissions in per cent, from eight sectors, (power houses, industries, transport vehicles, agricultural byproducts, fossil fuel recycling, waste disposal and

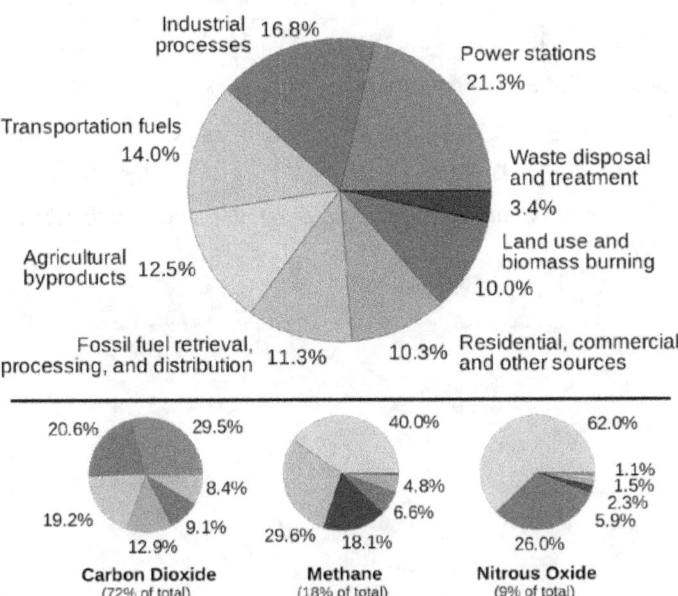

Fig. 2.3, Sector-wise annual generation and emission of greenhouse gases [coloured version of the fig is on page 234]

treatment, land use and bio-mass burning and resident, commercial and Misc. Sources) measured from the year 2000 are given in per cent.

Increase of Greenhouse Gases

The increase in the concentrations of many of the greenhouse gases was only 50 ppm for 200 years, from 1770 to around 1973. Another increase of 50 ppm took only 33 years, i.e. to 2006; from pre-industrialisation level of 280 ppm to 380 ppm in 2006. Atmospheric Chemistry Observational Databases are available on internet. The greenhouse gases with the largest radiative forcing are:

Gas	Current (1998) Amount by volume	Increase over pre-industrial (1750)	Percentage increase	Radiative forcing (W/m²)
Carbon dioxide	365 ppm {383 ppm (2007)}	87 ppm {105 ppm (2007)}	31% {37.77% (2007)}	1.46 {~1.532 (2007.01)}
Methane	1,745 ppb	1,045 ppb	150%	0.48
Nitrous oxide	314 ppb	44 ppb	16%	0.15

The contribution, of heat, by CFCs is negligible

(Source: IPCC radiative forcing report 1994 updated to 1998.

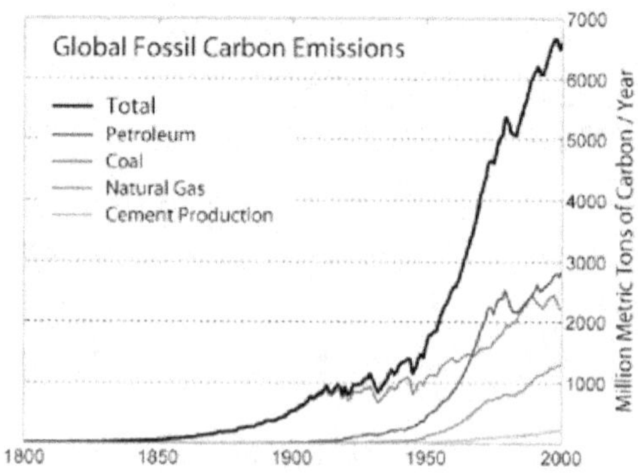

Fig. 2.4, Global carbon dioxide emissions 1751–2000
[coloured version of the fig is on page 235]

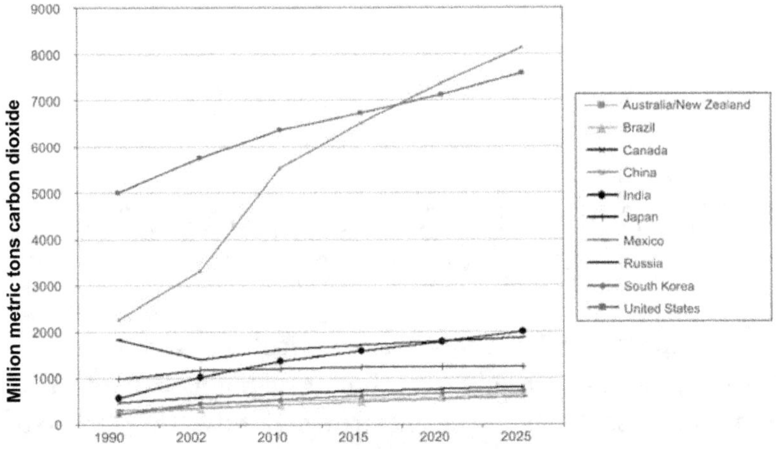

Fig. 2.5, Historical and projected CO_2 emissions country wise
(Source: Energy information administration)
[coloured version of the fig is on page 235]

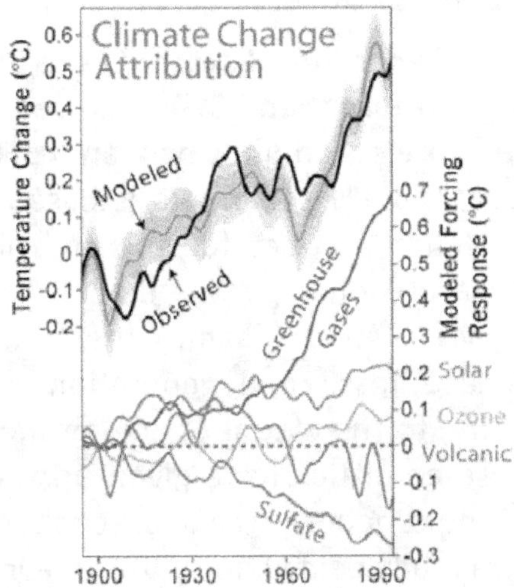

Fig. 2.6, Attribution of recent climate change
[coloured version of the fig is on page 236]

3

ENVIRONMENTAL DEGRADATION BY HUMAN LIFE STYLES

IPCC reports claim that industrialisation and resulting 'life styles' of humans are the principal cause of GHG emissions. Sector wise these can be divided as (a) Electricity Generation 31%, (b) Transport Sector 27%, (c) Industrial Sector 21% (d) Commercial and Residential Sector 12% (e) Agriculture Sector 9%. These distributions can vary country-wise, but the power sector remains universally the highest contributor of GHGs.

1) Electricity Generation Sector: CO_2 is the abundant part of the GHGs (more than 70%), followed by Methane. Nitrous oxide is released in small and SF6 in minute quantities (less than 1%). This data is based on coal, gas or other petroleum product fuels. Increasing efficiency by 'super critical' technology has helped to reduce CO_2 emission per MW. All the same, there are strenuous efforts to replace fossil fuel generation by renewable source electricity, to the extent of 50% or more. Denmark and a few more countries have given up the fossil fuels completely. Simultaneously, cleaner coals, (rectified or washed) are gradually being adopted. There is a worldwide aim to cut out their use, for electricity, by 2060 or earlier.

Many thermal stations are trapping their CO_2 and sequester it deep into soil. Methane is collected to be converted into 'Methanol' for myriad uses. Water is collected cooled and treated to remove all pollutants which can be safely released into streams or rivers. Wastes and garbage of all kinds are recycled or converted into compost for local utilisation.

But there is no denying that the GHGs, from thermal power stations, collected over a hundred years, since year 1900, by the rich countries, America and Europe's countries, cannot be wished away lightly.

2) Transport Sector: Petrol and diesel run vehicles proliferated from early 20th century. USA, Europe, Japan, South Korea and now India are turning out such vehicles in millions. Their numbers in cities are excessive literally, in numbers and pollution.

Studies in many cities in industrialised or metros in 'Developing' countries, have found that increasing vehicle emissions contribute to warmer air. Because, a vehicle emits mainly four pollutants; hydrocarbons, nitrogen oxide, carbon monoxide and matter particles. Carbon-monoxide becomes carbon dioxide which consumes OH particles in air, leading to increase methane in the atmosphere. Hydrocarbon and NOX leads to regional ozone and hemispherical imbalance. Diesel vehicles emit matter which turns intro black carbon. All of these pollutants, leads firstly to local warming, then to regional warming and finally to global warming.

The vehicular pollution is being curtailed through improved type of diesel, or using 'Biodiesel'. Type VI compliant vehicles are in operation in Europe and USA. India, too, is moving faster than before to achieve that level.

Biodiesel is one of the most efficient fuels; for every unit of fossil energy it takes to make biodiesel, 3.2 units of energy are gained. A US Department of Energy study has found that biodiesel, in comparison to petroleum diesel produces 78.5% less CO_2 emissions. Biodiesel is considered biodegradable, under ideal conditions is non-toxic. India has made it mandatory to sell, 5% biodiesel blended with

95% petroleum diesel in Delhi, Visakhapatnam, Haldia and Vijayawada. (India Today December 14, 2015)

Besides, the above, a number of measures to obviate the woes of auto-sector, are suggested in Chapter 4.

3) Industry Sector: Without getting into the complexities of methods of production of metal industries e.g. metals, (steel, aluminum, copper lead etc.), cement making and chemical industries etc., can be summarised that most of them have a) captive power generation, b) furnaces and heat based processes (blast furnaces, limekiln, boilers etc.), c) use of electricity like any commercial undertaking, d) chemical processes and effluents and finally e) direct release of CO_2 from mines, coal mines in particular.

History of industry reminds that the developed countries' industrial towns were soaked with soot given out by coal burning chimneys; (Charles Dickens novels of full chimney cleaning blackened chimney boys). Nearly 150 years of such industries, in Europe, Great Britain and America amassed millions of tons of CO_2 emissions in the atmosphere.

All in all, steel and cement plants contribute six and five percent respectively of GHGs in developed countries. And these do not include the emissions of users steel, cement and all the productions of a pile of industries in the world of today.

4) Residential and Commercial Activities: 12% of the GHGs in USA result from the residential and commercial buildings and their occupants, as detailed below:-

a) Natural gas, petroleum products and fossil fuels used for cooking and heating primarily (Coal is of limited use though coal and hay are more prominent fuels in less developed countries). Carbon dioxide-CO_2, methane-CH_4 and nitrous oxide-N_2O are the major expellants.

b) Waste landfills emit methane,

c) Wastewater treatment plants release methane and nitrous oxide and

d) Refrigerants (HFCs) are leaked from home and commercial air conditioners.

5) Agriculture, Horticulture and Forests: Surprising though, the foods and fruits that humans consume cause harm to our ecology. Forests too emit. GHGs are released to the extent of 9 percent of the global total of these emissions; 6 billion tons in 2011, which are likely to grow to 7 billion tons by 2030. One of the lacuna in these data is that large sections of the agriculture and horticulture sector do not know the methods of metering and recording the emissions. India has too many small holdings to create the desired information.

We do know that trees alternately release and absorb CO_2. But mostly, CO_2 gets emitted from the soil (treated with chemical fertilisers), biomass and smoke when haystacks and wastes are burnt. In addition, this sector throws up large volumes of methane, nitrous and nitrogen oxides and carbon-oxide. Manures and composts release methane and nitrous oxide.

China, Brazil, USA, India and the Russian Federation are in the league of largest GHGs contributors. Asia and sub-Saharan Africa will cause bigger additions to emissions in the first half of the 21st century; because of growth in populations and there for greater production and consumption of foods, vegetable oils and animal products.

The measures to curtail this sectors GHGs are:

i) About 24% of foods are wasted a US family throws away 1600USD worth foods annually. Cut down wastages by preservation processes and better eating habits.

ii) Reduce consumption of animal foods. These products cause huge wastes of water and energy, lead to toxic effluents.

iii) Better fertiliser management and alternate cropping techniques. The same applies to grazing lands.

iv) Farmers must learn to understand the types of emissions and how to manage them. They should obtain guidance from their respective governments. Measuring correctly the GHGs and reporting the data to the designated authorities is a must. The correct magnitude of the problem is lacking pres.

Note: The per cents and tons of GHGs relate to USA only. The relative emissions for India, and Europe remain similar. Mitigation measures are more elaborately given in the next chapter. The researches in another chapter could have become more adaptive, since these are reported which can be of about two to five years earlier progress. Solar Power applications are gaining ground at lesser costs.

Environmental Degradation by Industrial Effluents

POLLUTION because of human activities goes beyond emission of GHGs. Pollutants generate many more injurious agents. These effluents, besides causing climate change, are the primary cause of degradation of environment affecting health of humans, numerous diseases, also impact plants and animals. In the following pages, impact of pollutions, other than CO_2, is described.

Kinds of Pollution: (i) Air (atmospheric and indoor) (ii) Water, (iii) Sewage, (iv) Solid waste (v) Industrial waste (vi) Pesticides, (vii) Oil Spills (viii) Chemical fertilisers and composts (including animal manure) and (ix) Fertilisers and animal manure.

(i) Air Pollution

WHO believe about one fifth of the world population, mostly in developing countries is exposed to hazardous levels of air pollution, described below –

(a) **Outdoor Air Pollution:** Hundreds of millions of tons of inhospitable gases, vapour and solid particles rise into atmosphere from houses, vehicles, factories, power stations and forest fires. Fermentation over solid and sewerage wastes add harmful gases. Businesses such as dry cleaning throw up perchloroethylene, a pollutant, into air.

Smog (a brown, hazy mixture of gases and particles) reacts with sunlight and creates a host of harmful chemicals. It also forms toxic ozone. In concentrated quantities, these causes headache, irritation of eyes, damage the respiratory tracts of humans, damage plant life and even destroy trees.

Coal, gas, and oil burnt at power plants and inside automobiles, leave sulphur and nitrogen oxides in the atmosphere. These adulterants react with water in vapour and results in acid rain. High fall of acid rain can affect forests and contaminate soil as has been recorded in North America, Scandinavia, and central Europe.

One of the chemicals, CFC, (used in refrigerators and air conditioning plants) rises into higher levels of atmosphere and destroys ozone gases. The ultra violet radiations from sunlight pierce through the thinned ozone layer and reach our earth. Radiations destroy plants and may also cause skin cancer. Hence, the use of CFC has been banned universally.

(b) **Indoor Air Pollution:** Indoor pollutants are tobacco smoke, gases from ovens and furnaces, plus chemicals in households and fiber particles. Paints give off harmful fumes. Residents living in polluted surroundings suffer from headaches, eye irritation etc. Many people develop issues such as high blood pressure, mental stress, and sleeping disorder.

(c) **Ozone as a Pollutant:** An isotope of oxygen, Ozone with three atoms oxygen, is formed by a photochemical reaction, with other gases e.g. nitrous oxide with hydrocarbons (released prominently by auto-vehicles) which requires high intensity light and heat. Inhaling this gas more than an hour is injurious. The ozone content goes up in months of April to June. This gas hurts lungs leading to chest pains, asthma and bronchitis.

(ii) Water

Water pollution is extremely damaging. It has many causes and characteristics. High density nutrients decompose plants and absorb too much oxygen. Sewage, if not treated, leads to oxygen depletion in the receiving water. Industries discharge a variety of pollutants in their wastewater including heavy metals, organic toxins, oils, nutrients, and solids. Discharges can also have thermal effects, especially those from power stations. Silt bearing runoff from construction sites, deforestation and agriculture can inhibit the penetration of sunlight through the water column, restricting photosynthesis and causing blanketing of the lake or river bed. As a result, the polluted waters carry a variety of chemicals, all harmful. Pathogens produce waterborne diseases in both human and animal hosts. Alteration of water's physical chemistry, include acidity, conductivity, temperature increase, and destroys oxygen. The municipal water supplies even in Delhi, Mumbai, Bangaluru, Chennai, Kolkata and other large cities, may present health hazards. Water pollution, in the global context, is believed the leading cause of deaths and diseases, possibly for more than 14,000 mortalities daily.

(iii) Sewage

Piped sewage services exist, only in major towns, and these leave out slums and Jugghi Jhopris. Sewage treatment, and recycling facilities are old or inefficient. Human wastes may also be collected from open drains or even carried away in vehicles to distant places. The collected waste may be disposed into flowing waters and stored on landfills which are surely damaging to the environment. Cities, smaller ones in particular, have only open drains carrying soaped washing waters, laundry wastes and similar polluting agents. Any of the effluents, if not safely disposed, humans, animals and the environment suffer.

Treated sewerage has nitrates and phosphates which cause growth of algae in water. Bacteria in water use up algae, reduce oxygen content and thus kill aquatic beings. Untreated sewerage contains bacteria which when pollutes potable water, can cause cholera, dysentery and other water-borne diseases. With rising populations, these two kinds of pollution can reach menacing levels.

(iv) Solid Waste

Solid wastes come from households, agriculture and mining operations. Industrial operations cause organic solids besides plastics, glass bottles, paper, cans, and garden and crop wastes. These are dumped in open land-fills or thrown into rivers. Toxins in them either seep into soil or pollute the water. If burnt then the smokes pollute the air and may release metal particles in the environment. GHGs gathered in the atmosphere.

(v) Industrial Waste

It contains harmful chemicals and particles of metals particularly mercury and lead. Mixed with air they cause respiratory problems. These toxic particles can be absorbed by animals whose meat or milk can get into food consumed by humans, leading to various diseases, including cancer. Food factories processing meat, fish and fowl need huge quantities of water to cleanse the killed animals and birds. Guts and blood of animals also are thrown into rivers. Diseased effluents and toxic preservatives are killers of water-borne species; thus, adding another type of harmful ingredient. Warm and hot effluents are also killers of aquatic species.

(vi) Pesticides

Modern cropping of fruits and cereals absorb large quantities of pesticides. Besides killing even useful bacteria, these chemicals kill algae and reduce oxygen killing aquatic species. Pesticides also get consumed by humans, leading to hosts of health problems. US researches show that even healthier fruits and vegetables contain remnants of pesticides, with dangerous consequences. Pesticides which get mixed with milk, can infuse diseases, lethal may be, for infants and young children. Also pests that die can mix with principal crops; a hazard never imagined or taken care of!

(vii) Oil Spills

Petroleum and gases when spill over in ocean waters, they are health hazards and destroy aquatic species. Birds pecking on over-spills die without respiration, in large numbers. Gases oozing out, pollute atmosphere. And chemicals used to remove oil spills can lead to more dangers.

(viii) Chemical fertilisers and composts (including animal manure)

Fertilisers generally enhance productivity of agricultural and horticultural crops. Chemical fertilisers contain nitrogen, potassium and phosphorous plus copper, zinc etc., in micro quantities. Compound fertilisers have been developed to obtain superior outputs of both agricultural and horticultural applications. Pesticides are strong chemicals. These kill pests which otherwise would have destroyed crops, or at least reduced output.

More than 300 million pounds of different chemical poison are now produced in the form of fertilisers and pesticides under different brand names (Tomkins & Bird, 2002). The high solubility of such fertilisers also exacerbates their tendency to degrade ecosystems. Storage and application of some fertilisers, in some weather or soil conditions, can cause emissions of the greenhouse gases nitrous oxide (N_2O) and Ammonia gas (NH_3). Besides supplying nitrogen, ammonia can also increase soil acidity (lower pH, or "souring"). Excessive use of synthetic fertilisers and pesticides has caused tremendous harm to the environment as well affected human population indirectly. Continuous use also results in developing resistance of the pest, which become difficult to control by other means. The use of synthetic chemical fertilisers leads to imperfectly synthesized protein in leaves, which is responsible for poor crops and in turn for pathological conditions in humans and animals fed with such deficient food (Talukdar et.al.2003).

Researchers have established residues of chemicals in human diet and are of major concern today. For example "Pesticide Residues" in mg/day/person, country-wise are shown as below:

USA	7.6	UK	12.0
Canada	13.3	Australia	20.0
Germany	149.0	Europe	156.0

While India Non-vegetarians 356.3, Vegetarians 362.5

(Source: Organic horticulture, TNAU)

Indians take nearly 35 times more pesticides through food items than the average Americans.

A good portion of water soluble nitrogen fertilisers is lost to the ground water through leaching or run off. The excess nitrate leached into rivers or ponds encourages the growth of organisms and thus, a lot of organic matter produced which on decomposition lead to bad smell. This exaltation has an adverse effect on health. Foods grown with chemical Fertilisers cause various deteriorating health hazards in animals as well as human beings, for example:

(a) herbicides affect the central nervous system, respiratory and gastro intestinal system of human beings,

(b) some pesticides can also cause wheezing and nausea by irritating the lungs if large amounts are inhaled,

(c) chemical residues also cause depression, insomnia, oral acetomatism, myoclonus and hyper reflexia of man,

(d) accumulation of excess nitrogen in plants causes an infant disease, methaemoglobinemi

(e) amines produced from the nitrogenous fertiliser may cause cancer in human beings

(f) aluminum at high levels leads to birth defects, asthama, alzeimers and bone diseases. and

(g) calcium toxicity results in developmental and neurological toxicity, growth retardation, cognitive delay, kidney, nervous and immune system damage. And much more damages.

Conclusion

Humans are responsible for increase in GHG emissions. Usage of chemicals based production of foods and fruits, refrigeration and preservation chemically, and processed foods and drinks, all are injurious to soils, plants and humans. There are growing health hazards, in fact serious ailments. Environment is getting polluted causing serious damages to ecology.

———

4

A HISTORICAL PERSPECTIVE OF UNFCCC, IPCC AND THEIR CONTRIBUTIONS

United Nations Framework Convention on Climate Change

The genesis of a body under the aegis of United Nations was laid at the Earth Summit at Rio, Brazil in June 1992. An Intergovernmental Negotiating Committee had presented a report containing  text of the Framework Convention, which was duly adopted at the Rio Summit. 154 Parties (nations) signed a treaty in March 1994, to form a body called United Nations Framework convention on Climate Change. The signatory nations rose to 196 by 2014.

Among the immediate objectives of UNFCCC was to get GHG emission inventories of the developed nations. These data helped to establish the 1990 emissions as the 'benchmark' for all decisions, relating to reduction of emissions, by Developed Nations, named Annexure-I countries in the Kyoto Protocol formalised in 1997. The Framework Convention continues to lead efforts to bring down GHG emissions until the 'world ecosystems' are stabilised to halt Climate Change naturally, food production is adequate to feed all humans on our earth, and every nation can proceed with its 'economic development' without affecting ecology.

Since 1995, annual 'conferences' (meetings) have been held in Conferences of Parties (COPs- Representatives of Member States) to monitor progress in mitigation and adaption measures by members.

The Objective

In 1997 delegates of 190 countries agreed to Kyoto Protocol. But only 150 nations ratified this Agreement. USA, Australia, Canada, Russia, Japan and some other nations did not ratify Kyoto Protocol. Because, the Agreement required Developed Nations to accept its stipulations on them as legally binding; the rich nations declined to accept this condition. Given below are only important decisions at some of the COPs.

i) Bali Action Plan 2007

UNFCCC secretariat reported that 42 developed countries, 57 developing countries and the African Group had submitted their mitigation targets. The proceedings reported that developing country Parties agreed to "Nationally Appropriate National Mitigation Actions (NAMAs) context of sustainable developments supported and enabled by technology, financing and capacity-building, in a measurable, reportable and verifiable manner".

ii) Copenhagen Accord 2009

Heads of States had to rescue this very contentious meet. The Kyoto Protocol I was due to expire in 2012. Kyoto Protocol Annexure II countries were adamant at their CDR rights seeking compensation from 'rich' countries. The latter were in no mood to accept their demands.

The Heads of States prevailed to arrive at the Copenhagen Accord. The Accord states that global warming should be limited to below 2.0 °C (3.6 °F). This may be strengthened in 2015 with a target to limit warming to below 1.5 °C. The Accord does not specify what the baseline is for these temperature targets (e.g., relative to pre-industrial or 1990 temperatures). According to the UNFCCC, these targets are relative to pre-industrial temperatures, 114 countries have agreed to the Accord.

The UNFCCC secretariat noted that "Some Parties" informed the secretariat their "specific understandings on the nature of the Accord and related matters, based on which they have agreed to the Accord." The Accord was not formally adopted by the Conference of the Parties. Instead, the COP "took note of the Copenhagen Accord."

As part of the Accord, 17 developed country Parties and the EU-27 have submitted mitigation targets, as have 45 developing country Parties. Some developing country Parties have noted the need for international support in their plans.

iii) Cancun COP 2010

Many aspects of the Copenhagen Accord were brought into the formal UNFCCC process, as part of the Cancun agreements adopted by the COP in 2010. The agreement states that global warming should be limited to below 2.0 °C (3.6 °F) relative to the "pre-industrial level". This target may be strengthened "on the basis of the best available scientific knowledge, including in relation to a global average temperature rise of 1.5 °C".

As part of the Cancun agreements, developed and developing countries submitted mitigation plans to UNFCCC. These plans are compiled with those made as part of the Bali Action Plan.

iv) Durban and Doha-2011 & 2012

In 2011, India insisted on CDR principles enshrined in the Kyoto Protocol. The 'rich' declined to accept the stipulated requirements of 1997. However, the COPs adopted the "Durban Platform for Enhanced Action". As part of the

Durban Platform, the Parties agreed to "develop a protocol, another legal instrument or an agreed outcome with legal force under the Convention applicable to all parties".

2015 was set the 'target date' to arrive at the "New Treaty" at its 21st COP, which should come into force from 2020. Thus, the stage was set for Paris Treaty. As a corollary, the Kyoto Protocol was extended to 2020.

At Durban and Doha, Parties noted "with grave concern" that current efforts to hold global warming to below 2 or 1.5°C relative to the pre-industrial level appeared inadequate. The economic development needs of developing country parties were reiterated as follows "the Durban Platform reaffirms that social and economic development and poverty eradication are the first and overriding priorities of developing country Parties, and that a low-emission development strategy is central to sustainable development, and that the share of global emissions originating in developing countries will grow to meet their social and development needs".

UNFCCC informed the Parties about its prediction of temperatures during 2000 -2100 years and the impact on Climate Change and the disastrous consequences on human societies and other species. UNFCCC recommended the following:

i) To stabilise atmospheric GHG concentrations, global anthropogenic GHG emissions would need to peak, between 2060-2080, and then decline.

ii) Lower stabilisation levels would require emissions to peak and decline earlier compared to higher stabilisation levels.

iii) Members were made aware of the projected changes in annual global warming in 21th century for a range of emission scenarios.

iv) There is uncertainty over how GHG concentrations and global temperatures will change in response to anthropogenic emissions

The Intergovernmental Panel on Climate Change (IPCC) Established 1988

It was a recommendation by World Meteorological Organisation (WMO) and United Nations Environment Programme (UNEP). It is a scientific intergovernmental body, endorsed by the United Nations General Assembly. Membership of the IPCC is open to all, 196 odd, governments. The IPCC produces reports, which enable delegates of the UNFCCC to frame policies and rules which, are purported, to all nations to support the United Nations Framework Convention on Climate Change. IPCC reports cover "the scientific, technical and socio-economic information relevant to understanding the scientific basis of risk of human-induced climate change, its potential impacts and options for adaptation and mitigation." Members of IPCC, nominated by respective governments, are generally experts of scientific bodies in their country.

The IPCC does not carry out its own original research, nor does it do the work of monitoring climate or related phenomena itself. The IPCC bases its assessment on the published literature, which includes peer-reviewed and non-peer-reviewed sources. Thousands of scientists and other experts contribute (on a voluntary basis, without payment from the IPCC) to writing and reviewing reports, which are then reviewed by governments. IPCC reports contain a "Summary for Policymakers", which is subject to line-by-line approval by delegates from all participating governments. Typically this involves the governments of more than 120 countries. *Inputs by IPCC, was a major contributor to the member countries signing KYOTO PROCOL in 1997 . More than that, the 2007 Nobel Peace Prize was shared, in equal parts, between Mr. Al Gore. former Vice-President of the USA, and IPCC (Received by Mr. R.K. Pachauri, of India).*

IPCC has produced Annual Reports on Global Warming and its impact on climates, since 1991. However, its Report IV of 2007 and Report V of 2014 are of seminal importance. For these Reports, IPCC constituted Working Groups, generally known, WG I, WG II, and WG III; submitted one report each. In addition, Synthesis Reports were developed by another set of renowned scientists of sciences and meteorologists (hundreds of them).

Following are the areas of study of the three Working Groups:

i) WG I: Climate Change 2013, "The Physical Science Basis".

ii) WG II: Climate Change 2014, "Impacts, Adaptation, and Vulnerability".

iii) WG III: Climate Change 2014, "Mitigation of Climate Change" .

Each Report included Summary for Policy Makers, Technical Summary and Frequently asked questions.

iv) A Synthesis Report: This Report integrates the findings of the WG I, WG II, and WG III. The SYR 2014, also included findings of two Special Reports on (a) Renewable Energy Sources and Climate Change Mitigation and (b) Managing the Risks of Extreme Events and Disasters to Advance Climate Change Adaptation (2011).

Assessment Report V (Extracts)

The final SYR was released on November 2, 2014 in Copenhagen. Policy Makers met in December 2014, in Lima, at the 20th Conference of Parties, under the aegis of UNFCCC, to prepare ground work for the final meeting of the Heads of States in December 2015, in Paris to AGREE, to the virtual Kyoto II (nominal term) on Climate Change to be operative from 2020 to 2030.

TOPICS OF SYR

1. Observed changes and their causes;

2. Future Climate Changes

3. Future pathways for adaptation, mitigation, and sustainable development

4. Adaptation and mitigation

1. Observed Changes and Causes

i) Warming of the climate system is unequivocal, and since the 1950s many of the observed changes are unprecedented over decades to millennia. The atmosphere and oceans have warmed, the amounts of snow and ice have diminished and sea level has risen.

ii) Anthropogenic greenhouse gas emissions have increased since the pre-industrial era, driven largely by economic and population growth, and are now higher than ever. This has led atmospheric concentration of carbon dioxide, methane and nitrous oxide that are unprecedented in at least 800,000 years. Their effects, together with other anthropogenic drivers, have been detected throughout the climate system and are extremely likely to have been the dominant cause of the observed warming since the mid 20th century.

The evidence for human influence on the climate system has grown since AR4. Human influence has been detected in warming of the atmosphere and the ocean, in changes in the global water cycle, in reduction in snow and ice, and in global mean sea level rise; and it is extremely likely to have the dominant cause of the observed warming since the mid 20th century.

iii) In recent decades, changes in climate have caused impacts on natural and human systems on all continents and across the oceans. Impacts are due to observed climate change, irrespective of its cause, indicating the sensitivity of natural and human systems to changing climate.

iv) Extreme Events- Changes in many extreme weather and climate events have been observed since about 1950. Some of these changes have been linked to human influences, including a decrease in cold temperature extremes, increase in warm temperature extremes, an extreme high sea levels and an increase in the number of heavy precipitation events in a number of regions.

Notes: number of cold days and nights have decreased, number of warm days and nights have increased, frequency of heat waves has increased in Europe, Asia and Australia. Floods, cyclones or draughts have increased.

2. Future Climate Changes

i) Future Climate Changes, Risks and Impacts - Continued emission of greenhouse gases will cause further warming and long-lasting changes in all components of climate system, increasing the likelihood of severe, pervasive and irreversible impacts for people and ecosystems. Limiting climate change require substantial and sustained reduction in greenhouse gas emissions which together with adaptation can limit change risks.

ii) Key Drivers of Future Climate Change - Cumulative emission of CO_2 largely determines global mean surface warming by the late 21st century and beyond. Projection of greenhouse gas emissions vary over a wide range depending on both socio-economic development and climate policy.

iii) Projected Changes in Climate System - Surface temperature is projected to rise over the 21st century under all emission scenarios. It is very likely that heat waves will occur more often and last longer and extreme precipitation events will become more intense and frequent in many regions. The oceans will continue to warm, and acidify, and global mean sea level to rise.

Steps Towards A Green World

iv) Future Risks and Impacts caused by changing climate -
Climate Change will amplify existing risks and create new
risks for natural and human systems. Risks are unevenly
distributed and are generally greater for disadvantaged
communities in countries at all levels of development.

3. Future Pathways for Adaptation, Mitigation and Sustainable Development

i) Adaptation and mitigation are complimentary strategies
for reducing and managing the risks of climate change.
Substantial emissions reductions over the next few
decades can reduce climate change risks in the
21st century and beyond can increase prospects for
effective adaptation, reduce the costs and challenge of
mitigation in the longer term and contribute to climate
resilient pathways for sustainable development.

ii) Effective decision making to limit climate change and its
effects can be informed by a wide range of analytical
approaches, ethical dimensions, equity, value judgments,
economic assessments and diverse perceptions and
responses to risk and uncertainty.

iii) Without additional mitigation efforts beyond those
in place today, and even with adaptation, warming by
the end of the 21st century, will lead to from high to
severe, wide-spread and irreversible impacts globally
(high confidence). Mitigation involves some level of co-
benefits and of risks due to adverse side effects, but
these risks do not involve the same possibility of severe,
wide spread, and irreversible impacts as risks from
climate change, increasing the benefits from near-term
mitigation benefits.

iv) Adaptation can reduce the risks of climate change
impacts, but there are limits to its effectiveness, especially
with greater magnitudes and rates of climate changes.
Taking a longer-term perspective, in the context of

sustainable development, increases the likelihood that more immediate adaptation actions will also embrace future options and preparedness.

v) There are multiple mitigation pathways that are likely to limit warming to below 2°C relative to pre-industrial levels. These pathways would require substantial emissions reductions over the next few decades and near zero CO_2 and other long-lived greenhouse gases by the end of the century. Implementing such reductions poses substantial technological, economic and social and institutional challenges, which increase with delays in additional mitigation and if key technologies are not available. Limiting warming to lower or higher level involves similar challenges but on different timescales.

4. Adaptation and Mitigation

i) Adaptation and Mitigation responses are underpinned by common enabling factors. These include effective institutions and governance, innovations and investments in environmentally sound technologies and infrastructures, sustainable livelihoods and behavioural and lifestyle choices.

ii) Adaption options exist in all sectors, but their context for implementation and potential to reduce climate-related risks differs across sectors and regions. Some adaptation responses involve significant co-benefits, synergies, trade-offs. Increasing climate change will increase challenges for many adaptation options.

iii) Mitigation options are available in every major sector. Mitigation can be most cost effective, if using an integrated approach that combines measures to reduce energy use and the greenhouse intensity of end-use sectors, de-carbonise energy supply, reduce net emissions and enhance carbon sinks in land-based sectors.

iv) Effective adaptation and mitigation responses will depend on policies and measures across multiple scales: international, regional, national and sub-national. Policies across all scales - supporting technology development, diffusion and transfer, as well as finance for responses to climate change, can complement and enhance the effectiveness of policies that directly promote adaptation and mitigation.

v) Climate change is a threat to sustainable development there are many opportunities to link mitigation and adaptation and pursuit of other societal objectives through integrated responses (high confidence). Successful implementation relies on relevant tools, suitable governance structures and enhanced capacity to respond (medium confidence).

Montreal Protocol (A Summary)

Chlorofluorocarbons (CFC) and its replacement Hydrofluorocarbons (HFCL) cause depletion of the "Ozone Layer" in the Stratosphere. The Vienna Convention, the UN Agreement accompanying the Montreal Protocol, signed in 1987, decreed phasing out of these gases from 1991 to 2010. "Perhaps the single most successful agreement to-date has been the Montreal Protocol", declared Kofi Annan, UN Secretary General in 2003. Depletion of the 'Ozone layer' would have caused higher temperatures and many health hazards for humans, animals and plants. To date the Agreement has averted release of 135 billion tons of CO_2 emissions. Montreal Treaty, is well structured, supported and implemented sincerely by 196 nations. Its financial support to deserving nations is an example to be followed. (More detailed account in Chapter 7 of Part II). Montreal Treaty was amended in October 2016 to phase cut HFCs by 2050.

5

KYOTO PROTOCOL - ITS PRINCIPAL FEATURES

Country Wise Responsibilities

The Kyoto Protocol is an 'historic worldwide' agreement made in 1997, under the United Nations Framework Convention on Climate Change. The Protocol urged all nations to curb their emissions of Greenhouse Gases, particularly CO_2. There was a provision for countries to engage in emission trading; those who maintain or increase emissions of these gases they buy carbon credits from those who cut down GHGs more than their committed obligations.

The Kyoto Protocol is an amendment to the United Nations Framework Convention on Climate Change

Kyoto Protocol	
Opened for signature	December 11, 1997
Conditions for entry into force	Min. 55 parties of UNFCCC Annex I parties, agree to reduce at least 55% CO_2 of 1990 emissions.
Entered into force	February 16, 2005
Status, as of December 2006	169 countries and other governmental bodies agreed to reduction upto 61.6% CO_2

The treaty was negotiated/ finalised by representatives of nearly 190 countries in Kyoto, (near Oklahama) Japan in December 1997. The treaty was opened for signatures on March 16, 1998, and closed on March 15, 1999. It was ratified by 55 % of the countries, emitting more than 55% of GHGs, in 2005. The Protocol was to last till 2012. (subsequently extended until 2020) The Kyoto Protocol when ratified binds that body to be governed by global legislation enacted under the aegis of UNFCCC.

1. Objectives of Kyoto Protocol

i) Kyoto Protocol is intended to cut global emissions of greenhouse gases, and bring down their concentrations in the atmosphere at a level that would prevent dangerous anthropogenic (pollution originating in human activity) interference with the climate system.

ii) 1990 was designated the base year. All reductions in GHG emissions are to be counted with respect to corresponding conditions prevailing then, in the designated country.

iii) Temperature riser be limited to 2°C over and above those prevailing prior to the Industrial Revolution.

2. Details of the Agreement

According to a press release from UNFCCC and the United Nations Environment Programme the world's nations were divided into 2 groups listed in Annexure I and Annexure II:

i) Annexure I includes the Developed and Rich Nations (nearly 38 countries). These are USA, countries in European Union, including the erstwhile Eastern Europe, Russia, Japan, Canada, Iceland and Australia. Annexure I countries were deemed to be responsible for :

    a) the accumulated GHGs, since the Industrial Revolution.

    b) The Kyoto Protocol is an agreement under which industrialised countries will reduce their collective emissions of greenhouse gases by 5.2% compared to the year 1990 (but note that, compared to the emissions levels that would be expected by 2010 without the Protocol, this limitation represents a 29% cut). The goal is to lower overall emissions of six greenhouse gases - carbon dioxide, methane, nitrous oxide, sulfur hexafluoride, HFCs, and PFCs - calculated as an average over the five-year period of 2008-12. National limitations range from 8% reductions for the European Union and some others to 7% for the US, 6% for Japan, 0% for Russia, and permitted increases of 8% for Australia and 10% for Iceland.

c) Annexure-I countries were required to compensate the poor and developing countries with Financial and Technology Transfers to enable them keep emissions under control while proceeding with their economic development programmes.

USA, Canada Australia, Japan and Russia did not ratify Kyoto Protocol; Russia signed the Protocol later. These nations decided to independently manage their, respective, environment sustenance programmes.

ii) Annex-II countries lists mostly, developing countries nearly 161 countries in all. The prominent among these are China, India, Brazil, South Africa, Egypt, Iraq, Syria, Bangladesh, Pakistan, etc., group of countries in Africa and low lying island countries.

a) Annexure-II countries were allowed common but differentiated responsibilities, better known as CDR because per capita emissions in developing countries are still relatively low; and the share of global emissions originating in developing countries will grow to meet their social and development needs.

b) Annexure-II countries were to take up Clean Development Mechanism, CDM, projects which besides reducing GHG emissions will usher in better technologies for better adaptive measures.

iii) By February 2006, 169 countries ratified Kyoto Protocol.

Representatives of Australia and the United States on IPCC signed the protocol but their governments refused to ratify it. They did not accept responsibility of capping their CO_2 emissions, as per the Protocol. They followed their own programmes. The treaty, as already stated, was to run until 2012.

3. Enforcement of Kyoto Protocol

Kyoto Protocol stipulated a penalty on failed states. For every ton of GHG emissions above its agreed limit, the failed state ought to earn 1.3 times the emission shortage in the next commitment period,(yet to be decided). Such countries, as an alternative, were allowed to buy "Carbon Credits" described below.

4. Carbon Trading and Carbon Credit System

The Protocol framed an ingenious mechanism of 'Carbon Trading'. The defaulter country could buy Carbon Credits (CC) from a developing country which earned the CC by successfully completing projects under Clean Development Mechanism (CDM). A CDM Board, based at Bonn, was established by UNFCCC to issue "Certified Emission Reductions" for buying and selling emissions.

Incentives were devised for Annexure II countries, to implement CDM projects with high efficiency, i.e. achieving lesser GHG emissions. They could earn profit from sale of Carbon Credits earned by them.

In addition to the above impetus, a Joint Implementation scheme, JI, was designed to help the former Soviet Union and countries in Eastern Europe to cross over its developing status to industrialised status at a fast pace.

Europe's Annexure-I countries have established Designated National Authorities to manage their GHG responsibilities. Japan, Canada, Italy, the Netherlands, Germany, France, Spain and Annex-II countries also set up Designated National Authorities to manage the Kyoto dictates. The latter specifically authorised those Authorities to plan and manage their CDM Projects and receive certificates from the CDM Executive Board to maximise the value of Carbon Trading rights.

5. Summary of Beneficial and Harmful Effects of Global Warming

There are, both, "beneficial" and 'harmful" effects of Climate Change. These are particularly assessed to work out the net "costs" of Climate Change (which possibly be compensated by annexure I countries).

5.1 Beneficial Effects

a) According to the Canadian Ice Service, the amount of ice in Canada's eastern Arctic Archipelago, decreased by 15% between 1969 and 2004. Iceland and similar areas will also release areas presently covered under ice-mountains for use by humans. Greater quantities of crops and transfer of populations will ease pressure on those governments', densely populated urban areas.

b) Melting Arctic ice may open the Northwest Passage in summer. Supertankers which are too big to sail through the Panama Canal and currently have to go around the tip of South America can use this route, saving 5,000 nautical miles (9,000 km) compared to the canal route.

c) Increasing carbon dioxide may increase ecosystems' productivity to a point. However, ecosystems' interactions with other aspects of climate change are not predictable. Therefore, likelihood of such improvement is not certain.

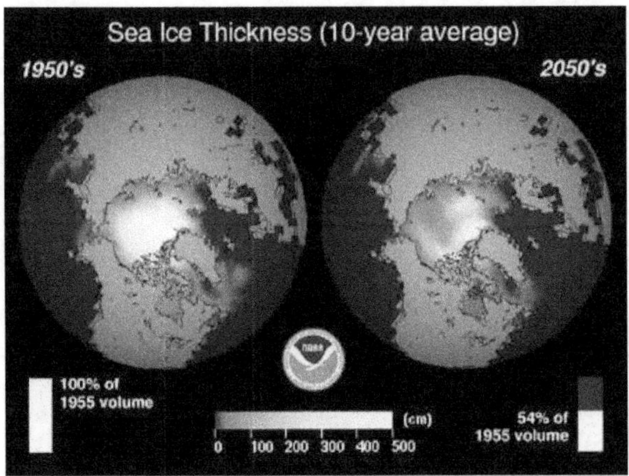

Fig. 5.1, Variation in sea ice thickness in Arctic-1950's - 2050's
[coloured version of the fig is on page 236]

5.2 Harmful Effects

While the reduction of summer ice in the Arctic may be a boon to shipping, this same phenomenon threatens the Arctic ecosystem, most notably polar bears which depend on ice floes. Subsistence hunters such as the "INUIT" peoples will find their livelihoods and cultures increasingly threatened as the ecosystem changes due to global warming.

Note: Chapter 6 gives a detailed account of 'estimates of all types of costs'.

6

KYOTO PROTOCOL
COSTS, BENEFITS OF REDUCING GHGs

The issue of costs was (is) inevitable. Initially, nations were cool; draughts and floods are "God's Will". What and Which costs? Answers were complex. A stronger resistance, came primarily from affluent societies, who will be made to defray huge expenditures. However, the scientific minded, tried to, firstly, identify "the costs" before assessing them.

i) The Rescue, Relief and Rehabilitation costs are the three basic components of assessing costs:

   a) Rescue efforts involve Disaster Management Agencies, Fire Brigades, Army, Navy and Air Force personnel to save human and animal lives, save bridges, railways and costly properties. Damage Costs (DC).

   b) Relief means providing shelters, beds and clothing, foods and drinks, medicines et al. This is known as Rehabilitation Relief Cost (RRC) and

   c) Rehabilitation costs can consist of restoring rail lines, communications, bridges, building new abodes for thousands, compensating agriculturists for lost crops, compensating businesses for loss of commercial properties et al. and therefore, its is called as Social Costs (SC).

   All in all, the expenditures on DC + RRC + SC could run into "lakhs of crores" spread over 3-5 years.

ii) Increase in Insurance and Maintenance costs.

iii) Mitigation costs (MC): These relate primarily to measures to curtail GHGs, e.g. adoption of better fossil fuel burning techniques, renewable sources of electricity generation and a lot more.

Insurance Risks and Costs

The insurance industry is hard hit by disasters in recent years. A threefold increase in major natural disasters since 1960 has increased insured losses fifteen fold in real terms (adjusted for inflation). According to 2002 ERM study, the climate changes caused 35–40% of the worst catastrophes from 1975 to 2001. In the same period, globally the affected population increased from 2% to 4%, the trend being virtually linear.

A June 2004 report by the Association of British Insurers, ABI, declared "Climate change was not a remote issue for future generations to deal with. It was, in various forms, here already, impacting on insurers' businesses right now". Risks for households and property were increasing by 2-4% per year due to changing weather. Claims in the UK for storm and flood damages were over £6 billion over 1998-2003, twice of the claims from 1993-1998. ABI stated in its report of 2005 that measures to reduce/ cap carbon emissions could avoid 80% of the projected additional annual cost of tropical cyclones by the 2080s. Rising claims mean higher insurance premiums which, ultimately, will become unaffordable for significant numbers.

Financial institutions, including the world's two largest insurance companies, Munich Re and Swiss Re, warned in a 2002 study that "the increasing frequency of severe climatic events, coupled with social trends" could cost almost US$150 billion each year in the next decade. These costs would raise premiums of insurance and steep rise in disaster relief which will burden customers, taxpayers, and the insurance industry alike. In the United States, according to Choi and Fisher (2003) each 1% increase in annual precipitation could enlarge catastrophe loss by as much as 2.8%. Gross increases mostly increased because of population and property values in vulnerable coastal areas, though there was also an increase in frequency of weather-related events like heavy rainfalls since the 1950s (Science, 284, 1943-1947).

Infrastructure Maintenance Costs

Roads, airport runways, railway lines and pipelines, (including oil pipelines, sewers, water mains etc.) may require increased maintenance and renewal as they become subject to greater temperature variation. Regions already adversely affected include areas of permafrost, which are subject to high levels of subsidence, resulting in buckling roads, sunken foundations, and severely cracked runways. These costs are hard to assess, with some firmness.

1. Mitigation Costs (MC)

A comprehensive view is given in the following chapter, "Carbon Trading". Suffice it to say here that at an IPCC conference in April 2007, delegates from 120 nations discussed the specific economic and societal costs of mitigating global warming. They formed a general consensus along the lines of the IPCC Fourth Assessment Report that benefits of mitigation of global warming are worth the mitigation costs.

2. Cost Estimates, Views and Controversies

i) Professor Robert O. Mendelsohn of Yale School of Forestry and Environmental Studies expressed doubts, at the 2004 Copenhagen Meet on 'Economic Effect of Global Warming'. He commented that. "The newer studies imply that impacts depend heavily upon initial temperatures (which vary according to latitude). Countries in the polar region are likely to receive large benefits from warming, countries in the mid-latitudes will at first benefit and only begin to be harmed if temperatures rise above 2.5°C. Only countries in the tropical and subtropical regions are likely to be harmed immediately by warming and be subject to the magnitudes of impacts". Summing up these regional impacts across the globe implied that warming benefits and damages were likely to offset each other until warming passes 2.5°C.

ii) An attempt was made to assess 'the Social Cost of Carbon' (SCC). Peer-reviewed estimates of the SCC for 2005 were (mean value of cost) US$43 per ton of carbon (tC); but the range around this mean is abnormally large. In a survey of 100 estimates, the values ran from US$-3/tC up to US$350/tC.

iii) The most acknowledged estimate is the Stern Review, a 2006 report by the former Chief Economist and Senior Vice-President of the World Bank, Nicholas Stern. He warned that climate change will have a serious impact on economic growth without "mitigation efforts". Mr. Stern inferred that an investment of one percent of global GDP is required to mitigate the effects of climate change. Failure to do so would mean risking a recession worth up to twenty percent of global GDP. The Stern Review was (and still is) criticised by some economists, on various counts; one being that he used an incorrect discount rate in his calculations. On the other hand, Stern had supporters too; that Stern's approach and estimates were reasonable, even if they employed their independent methods of calculating the estimates.

Summary

A report released on October 29, 2006 by Nicholas Stern stated that Climate Change could affect growth, which could be cut by 20%, unless drastic action is taken. (Report was a stark warning on climate).

A variety of studies show the summary of costs in the fig. 6.1 below.

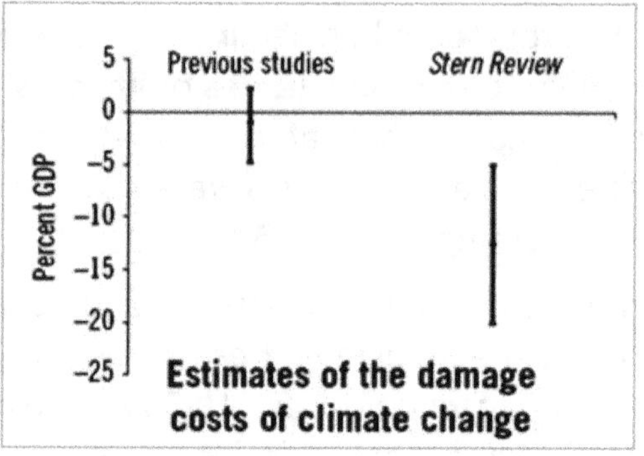

Fig. 6.1

Stern review saw no gains and much higher costs taking into account the socio-economic and bio-physical damages. Cost can vary on account of different assumptions of discount rates, duration and geographic location of the studied area. Consensus at IPCC, suggests MC will be lesser than the DC and RRC, besides sufferings caused by unsuspected, unexpected disasters.

7

KYOTO PROTOCOL - CARBON TRADING AND FINANCIAL COMPENSATION

UNFCCC and Kyoto Protocol policies' makers (COPs and their WGs) believed that a "price" of emissions must be determined to encourage mitigation efforts. The offenders must pay for failing to meet their commitments. But How? A plethora of issues required consideration, for example

i) Limits of Carbon Emission

ii) a) The Period for which costs, net of benefits, be recovered,

   b) Discount Rate, and

   c) Estimate of net Costs which be recovered.

iii) The method of Recovery

   a) Carbon Tax,

   b) Fine,

   c) Trading System; which requires determining the 'Unit' of emission and pricing mechanism and a lot more.

(i) Limits of Carbon Emission

One fundamental factor is the target (maximum) level of atmospheric carbon dioxide, which can be tolerated to contain the temperature rise set limit of 2°C. A common target level (assumed by the United Kingdom) is 550ppm (current levels are around 380ppm, and rising at 2-3ppm per year).

The Stern Review recommends adopting quantitative global stabilisation target range for the stock of greenhouse gases as a foundation for policy. It suggests that this target range would be likely to be somewhere between 450-550 ppm CO_2-e. It also recommends a carbon price signal through the use of a carbon tax or emissions trading scheme.

(ii) Discount Rate, Costs etc.

An IPCC TAR (Synthesis Report) suggested values of $78bn to $1141bn annual mitigation costs, amounting to 0.2% to 3.5% of current world GDP (which is around $35 trillion), or 0.3% to 4.5% of GDP if borne by the richest nations alone. As economic growth is expected to continue, the percentage would fall. In terms of cost per ton of carbon emission avoided, the range (for a target of 550ppm) is $18 to $80.

However, the Stern Review produced much larger benefit estimates, of between 5 per cent and 20 per cent of GDP. The difference reflected a number of factors, the most important of which were the choice of discount rate, the use of welfare weighting for effects on people in poor countries, a greater weight on damage to the natural environment and the use of more up-to-date scientific estimates of likely damage.

The rest of the 'pay back' determinants are, in totality, country specific i.e. the existing level of emissio, discount rate, period of 'pay back' etc. No discussion is presented.

(iii) Carbon Tax View

Advocates of mitigating climate change hold that greenhouse gas emissions must carry a price, so the market can price in the impact of their emission. This could take the form of a carbon tax or of emission caps, with a market created for trading emission permits, much as was done for sulfate emissions blamed for acid rain. Thus the economic impact of avoiding greenhouse gas emissions depends on how much consumption will have to be avoided, and how quickly the economy can incorporate efficiency gains.

"No regrets" policies - notably reducing fossil fuel subsidies, which are predicted to increase whilst reducing CO_2 emissions. Article 2 of the Kyoto Protocol specifies a progressive removal of subsidies and reform of taxes as a means of achieving reduction commitments.

McKibbin and Wilcoxen (2002) argue that a combination of long term carbon price signals and short terms caps on economic cost is needed to address economic efficiency, equity sharing and political feasibility. Brink et al (2005) showed that the costs of mitigation can be reduced by considering the inter-relationships of different greenhouse gases, and the differential impact that different technological decisions may have on their emissions.

Not many countries have so far adopted 'carbon tax' or 'subsidy' methods.

But UNFCCC and Kyoto Protocol allowed defaulters to buy " Certified Emission Reduction" Units, CERs", from the "good performers", mostly from among Developing nations. The exchange, "Selling-Buying" of CERs became known "Carbon Trading", which was preferred by rich and developed nations rather than pay hefty sums as compensation to developing countries.

Carbon Trading System

In essence, it is trading of CERs on open exchanges, which came up in various, advanced, countries. The institutions to certify CERs, and regulating their trading was retained for the institutions, set up and controlled by UNFCCC / Kyoto Protocol assisted by IPCC. These 'governing bodies', in addition, supported and encouraged, Non-Annexure I nations, to undertake "Clean Development Mechanisms", CDM, projects which were slated to usher in improved methods. The related country was assured of sustained economic development, besides, earning CERs. The CERs earned them finances to defray their expenditures on CDMs. (Explained below)

Clean Development Mechanism (CDM)

Note: All projects, and schemes are set by Kyoto Protocol; technical advice is provided by IPCC

The Clean Development Mechanism (CDM) are schemes and projects, undertaken in Non-Annexure I countries, to achieve reduction of GHG carbon emissions. The CO_2 or carbon reductions are measured in "emission reduction units". These are certified, by a UN approved body, as Certified Emission Reduction (CER) Units. The CERs are tradable at the open exchanges as listed in the section," Carbon Trading". The ultimate objective is towards achieving lowest carbon emissions to limit the global warming to 2°C. At the same time the Annexure I countries are helped to meet their responsibility of emission reduction. The defaulting countries buy CERs on the exchanges, and submit those in lieu of their deficient emissions.

The projects and the issuing of CERs is subject to approval, to ensure that these emission reductions are real and "additional." The CDM is supervised by the CDM Executive Board (CDM EB) and is under the guidance of the Conference of the Parties (COP/MOP) of the United Nations Framework Convention on Climate Change (UNFCCC).

2001 was the first year when CDM projects could be registered. Up to 7 September 2012, the CDM projects had earned, one billion CERs.

However, a number of weaknesses of the CDM have been identified (World Bank, 2010, p. 265-267). Several of these issues were addressed by the new Program of Activities (PoA) that moves to approving 'bundles' of projects instead of accrediting each project individually. In 2012, the report Climate change, carbon markets and the CDM: A call to action said governments urgently needed to address the future of the CDM. It suggested the CDM was in danger of collapse because of the low price of carbon and the failure of governments to guarantee its existence into the future. Writing on the website of the Climate & Development Knowledge Network, Yolanda Kakabadse, a member of the investigating panel for the report and founder of Fundacion Futuro Latinamericano, said a strong CDM is needed to support the political consensus essential for future climate progress. "Therefore we must do everything in our hands to keep it working," she said.

Carbon Trading

The position of 'Carbon Footprints, amount of CO_2, which could be traded was, in 2004, as per table below

(Source U.S. Environment Protection Agency-TOI 26-7-07)

Table 7.1, 10 Major CO_2 Producing Countries in 2004, in metric tons

USA	Canada	China	Russia	Japan	India	Germany	U.K.	Italy	S. Korea
5912	5880	4707	1685	1262	1112	862	580	485	497

All countries, generally, devolve their emissions targets to individual industrial entities, such as a power plant or paper factory. This means that the ultimate buyers of CERs are often individual companies that expect their emissions to exceed their quota (their Assigned Allocation Units). For shortages, typically, they purchase 'Credits' directly from another party through/ from a broker, e.g. from a JI/CDM developer, or on a trading exchange for CER Units.

National governments, some of whom may not have devolved responsibility for meeting Kyoto obligations to industry, and that have a net deficit of Allowances, will buy credits for their own account, mainly from JI/CDM developers. These deals are occasionally done directly through a national fund or agency, as in the case of the Dutch government's ERUPT programme, or via collective funds such as the World Bank's Prototype Carbon Fund (PCF). The PCF, for example, represents a consortium of six governments and 17 major utility and energy companies on whose behalf it purchases Credits.

Since CERs are tradable instruments with a market / transparent price, financial investors have started buying them for pure trading purposes. This market is expected to grow substantially, with banks, brokers, funds, arbitrageurs and private traders eventually participating. For example, "Emissions Trading PLC", was floated on the London Stock Exchange's AiM market in 2005 with the specific remit of investing in emissions instruments.

Although, Kyoto created a framework and a set of rules for a global carbon market, there are in practice several distinct schemes or markets in operation today, with varying degrees of linkages among them.

Kyoto enables a group of several Annex I countries to join together to create a so-called a cluster of countries that is given an overall emissions cap and is treated as a single entity for compliance purposes. The EU elected to be treated as such a group, and created the EU Emissions Trading Scheme (ETS) as a market-within-a-market. The ETS's currency is a EUA (EU Allowance). The scheme went into operation on 1 January 2005,

The UK established its own learning-by-doing voluntary scheme, the UK ETS, which ran from 2002 through 2006. This market will exist alongside the EU's scheme, and participants in the UK scheme have the option of applying to opt out of the first phase of the EU ETS, which lasts through 2007.

The Joint Implementation (JI) mechanism is another type of grouping, formed by a number of project developers. These JIs collectively contributed funds and perform carbon trading on behalf of their members. According to IETA, the market value of CDM/JI credits transacted in 2004 was EUR 245 m; it is estimated that more than EUR 620 m worth of credits were transacted in 2005.

Several non-Kyoto carbon markets are already in existence as well, and these are likely to grow in importance and numbers in the coming years. These include the New South Wales Greenhouse Gas Abatement Scheme, the Regional Greenhouse Gas Initiative (RGGI) in the United States, the Chicago Climate Exchange, the State of California's recent initiative to reduce emissions, the commitment of hundreds of US mayors to adopt Kyoto targets for their cities, and the State of Oregon's emissions abatement program.

Taken together, these initiatives point to a series of linked markets, rather than a single carbon market. The common theme across most of them is the adoption of market-based mechanisms centered on Carbon Credits that represent a reduction of CO_2 emissions. The fact that most of these initiatives have similar approaches to certifying their credits makes it conceivable that Carbon Credits in one market may in the long run be tradable in most other schemes. This would broaden the current carbon market far more than the current focus on the CDM/JI and EU ETS domains. An obvious precondition, however, is a realignment of penalties and fines to similar levels, since these create an effective ceiling for each market.

India's Efforts in Carbon Trading

Quite a few organisations have known of opportunities for Carbon Trading prompted by Kyoto Treaty. CDM projects were embarked upon since 2003, by a few companies. Grasim Cements earned Rs. 17 crores in 2006. JSW Steel successfully completed CDM projects worth 1.35 million CERs (carbon emission reductions) and traded them for Rs. 112 crores. JSW hopes to raise about 14 million CERs in the coming decade. JSW Energy group made 330 crores by selling 4 million CERS in the past 2 year. Gujarat Fluorochemicals, manufacturer of HFC-22 for refrigerators is destroying the HFC-23, a byproduct before it could leak to atmosphere. This procedure earned 6.5 million CERs last year and hopes similar gains annually from now on. The worldwide trading in carbon certificates and CDM projects is given in the table below:

Country	India	China	Brazil	Mexico	Chile	Malaysia	Republic of Korea	Others
CDM	35.32	13.76	13.76	11.9	2.51	2.12	1.85	18.78
CERs	14.95	44.23	10.72	3.94	1.97	1.17	8.81	14.21

All data are in per cent of the total. China undertakes CDM projects 5-7 times those of Indian organisations.

India has still to gain expertise and methodologies of the certification process, besides experience of efficient execution of CDMs. Moreover, it must be emphasised that Indian industry and manufacturers must adopt latest technologies to cut down emissions. "Merely burning GHG gasses and CDMs is a halfway house", says Sunita Narain, director, Centre for Science and Environment, an NGO at Delhi.

(Source: Business World, 27 August, 2007, "Running on Carbon" by Arundhuti Dasgupta).

Note: Currently, there are six exchanges trading in UNFCCC related to carbon credits: the Chicago Climate Exchange (until 2010[), European Climate Exchange, NASDAQ OMX Commodities Europe, Power Next, Commodity Exchange Bratislava and the European Energy Exchange. NASDAQ OMX Commodities Europe listed a contract to trade offsets generated by a CDM carbon project called Certified Emission Reductions. Many companies are now engaged in emissions abatement, offsetting, and sequestration programs to generate credits that can be sold on one of the exchanges. At least one private electronic market has been established in 2008: CantorCO2e. Carbon credits at Commodity Exchange Bratislava are traded at special platform called Carbon place. Trading in emission

permits is one of the fastest-growing segments in financial services in the City of London with a market estimated to be worth about €30 billion in 2007. Louis Redshaw, head of environmental markets at Barclays Capital, predicts that "Carbon will be the world's biggest commodity market, and it could become the world's biggest market overall".

Benefit Distribution

UNFCCC recommended that the financial benefits be distributed, to the following:

i) low-lying countries' who are under the risk of getting drowned in oceans

ii) countries subject to increased droughts are poor African countries

iii) ability of poor countries to mitigate / adapt (margin)

iv) GW increases variability of weather, which implies greater capital requirements for water storage systems, flood defenses, etc as well as individual requirements to cope with wider variation in weather patterns.

Cost Distribution Among Rich Countries

Just for illustrating, views for cost contribution, among many rich countries, are noted below. The 'Developed Nations' have evaded any responsibility on this count; instead, adopted their approach, independently.

i) The costs of mitigation may also be distributed unequally, both within and between countries. Wier et al (2005) showed that carbon taxes, in order to receive social acceptance be either within the environmental tax structure or in other parts of the tax system. Indirect taxes (on business) are less regressive, and petrol taxes are found to be progressive.

ii) Bastianoni et al (2004) note the differences between methodologies for assigning responsibility for greenhouse gas emissions, which include the geographical approach, based on the IPCC guidelines for GHG inventory; the consumer responsibility approach, based on the Ecological Footprint methodology; and the Carbon Emission Added (CEA) approach, which resembles the accounting of a Value Added Tax. Different methodologies can produce quite different results in terms of responsibility for emissions, with consequent impact on the policy.

———

8

MITIGATION AND ADAPTATION MEASURES
(For Individuals, Municipalities, Industries, Governments etc.)

Use of Electrical and Heat Energies and Water Installations

1. By Individuals and Families

Consumption of electricity by the domestic sector is about 16% of the consumption across all sectors. Some very simple habits and precautions in its use can result in lowering the generation of electricity thus reducing emission of GHGs at power stations. In a way the suggestions given below will support GOI's aggressive drive to contain global warming in the ensuing five years. These steps are:

i) Use LED bulbs, incandescent bulbs are likely to be banned. CFL bulbs also consume more energy and their dispensation after use are a problem. Buy energy efficient electric appliances certified by BIS with star ratings. An estimated annual savings of at least Rs.2000 is possible for a five star refrigerator or a geyser.

ii) Install solar water heaters and solar panels on roof tops. Their costs are coming down and there are government subsidies to popularise their use.

iii) Switch off electrical lights when natural light is available. Automatic switch off-switch on instruments are available

iv) New buildings be 'green buildings'. (See more in later pages) Glass panels in existing panels can also be coated with 'solar panels'

v) A controlled use of heat is also recommended by using pressure cookers and modern cooking vessels.

vi) (a) Eliminate leakage of water from taps and joints in pipes. Surprising it may be, that a continuously dripping tap, one drop per second, can amount to loss of 10,000 liters of water annually. A plumber can fix all leaks at nominal charges. Imagine saving of electricity from filtration to pumping all along the water pipes till your roof tops.

b) Recycle your bath room outflows in your gardens or for washing and mopping floors. In particular, discarded waters from the home-water filters are eminently usable for washing etc.

c) All households to install water harvesting systems for which ample support/advice is possible from architects and municipalities.

d) Grow vegetables in kitchen gardens or roof tops. Planting green plants is desirable.

2. Commercial and Corporate Establishments

i) All the measures given above, for electrical, heating and water systems, must be followed by 'Services Managements' of commercial buildings. Automatic switch on-switch off to suit daylight instruments and LED lighting systems are a must. Solar panels on roof tops and window-glass panels are recommended for the existing buildings, especially for 'malls'.

ii) Heating, Air Conditioning and Cooling systems must be most efficient; the older ones should be replaced. Much care is advised towards water pumps, piping and outlets and installing water harvesting installations.

iii) Leakages of heat, cooled air and moisture can be prevented by closely fitting/shut doors and windows.

iv) Grow greens as much as possible.

v) Green Buildings, henceforth, are essential for commercial and office uses.

3. Municipalities and RWAs

Besides what has been suggested here to fore, there are some special areas of attention for the above stated bodies:

i) Roads form a major areas of concern under their control. Their construction, and layout, allowing smooth flow of traffic, must ensure for fast draining of rain water especially during floods.

ii) Modern Drainage Systems are a must; should learn from Singapore, Japanese, US and European civic and coastal authorities. "Drainage Systems" are presently ill-designed and natural drainages are aborted (Mumbai and Chennai) for outflows. Pumping Systems prove inadequate whenever heavy rains happen, even during monsoons. Bombay roads get sub-merged against high tides. The open drains, as presently provided, must be thoroughly cleaned and emptied, before rainy seasons. Smart City Municipalities must provide efficient drainage systems. It is emphasised that the life of roads depends on fast outflow of waters, besides causing inconvenience to citizens and waste of fuels because of traffic jams.

iii) Sewage recycling to extract water for re-use, removal of toxics and disposal/recoveries from effluents are now practicable. Singapore is once again a worthy example .

iv) Garbage and household wastages' collection and disposal are becoming a big problem towards achieving "Swacch Bharat". Garbage dumping grounds are becoming a big nuisance besides hygienic dangers to citizens living nearby.

On the other hand, these discarded wastes are a "no loss" but "good profits" source. These materials when separated into organic materials can be converted into 'organic compost' for fields or 'bio-mass' to be used in households or electricity generation. Glass bottles,when crushed, can be made in to glass ware again. Plastic bottles and others are crushed to re-use for toys etc., and are now being used to reinforce for road building. Paper and cardboards have been recycled to produce papers and cardboards again. Paper-Machie is base for artistic products, specialised by artisans in Kashmir. Even electricity is being generated by recycling city wastes.

v) Municipal bodies or town planners can standarise designs for water harvesting systems for small residences to high rise buildings. Smaller abodes can be provided assistance in their construction and maintenance.

vi) A heavy responsibility lies on these Bodies to safeguard against depletion of water-table. Wherever possible water bodies be created or older ones preserved. Parks and city forests are desirable. For example, Delhi's Ridge forests is very useful for carbon dioxide absorbing and must be preserved. The city roads can be lined up with fruit or shade-providing trees.

4. Industries and Powerplants

The two together contribute more than half of the country's harmful gases:

Industrial sector is the biggest emitter of GHGs. Steel plants, cement manufacturing units and brick kilns are the principal offenders. Industries must employ the latest technologies at their plants.

Thermal, especially coal based, electric power stations emit very high quantities carbon dioxide and methane gases, both the principal inputs to temperature rise. There was a strong pressure, at the Paris Conference, to discontinue, by a fixed date, the use of coal for electricity generation. However, while resisting a fixed time schedule, the nations realised the benefits of discouraging use of coal. Even gas or fossil fuel based power generation must be given up in due course. France and Scandinavian countries have only 20% or even less thermal power plants.

Renewable sources must be used for power generation. Solar and Wind energy uses should be speedily adopted. Some countries have already achieved very high levels by such methods. India has plans to reach the target of 40% of the power generation by renewable sources by 2040. The coal based plants, in the mean while, will also be upgraded to be more efficient thus reducing GHG emissions.

Nuclear power is less polluting, but it is the costliest methodology. Moreover, disposal of nuclear wastes is a huge, risk prone, problem. Such plants take long times to become operational. Hence nuclear source for power is on the decline. Some countries are rapidly dismantling their nuclear plants, because of highly devastating accidents.

Transmission and distribution losses can be controlled/reduced by installation of specially designed equipment. India must follow practices in Sweden, Norway and USA.

Within their premises battery/electric vehicles be used. Of course, all the measures suggested for commercial, and domestic sectors must be utilised.

5. Central and State Governments

The Central Government carries the responsibility to devise policies to control Global Warming. Eight missions were set up previously to take care of international obligations. The Ministry of Environment is the 'nodal agency' to plan and oversee towards sustained ecological balance. It submitted India's INDC statement. The Minister-in-Charge participated in the negotiations at the Paris Conference. The State Governments are the ones directly responsible for implementing the programmes laid out by the Central Government. However, the respective areas of control and responsibility should be un-ambiguously demarcated. Some suggestions are:

i) Ensure compliance of the standards (to be mandatory) set for industries, car and vehicle manufacturers (besides, regular validation of those standards by car users), especially of heavy trucks and building machinery.

ii) Roads and carriageways be designed to enable free and fast movements. The level crossings of the railways be eliminated (Plans are underway).

iii) Invest in public transport systems, of buses and metros. Encourage private investments in this programme. Discourage one car one passenger users. Encourage use of electric and hybrid type vehicles

iv) Hopefully, Smart Cities will develop self contained districts of residential, commercial, local official establishments, amusement and entertainment places. Besides curtailing travelling times, there will be huge savings of fuels.

v) Levies/ fines be enforced on emitters of GHGs, more than stipulated standards .

vi) Promote 'carbon trading' among industrial undertakings, as well as transporters. The 'carbon trading' at international levels be encouraged.

vii) Encourage research in nano materials and their applications. (Note Research in latest technologies is described elsewhere).

viii) Ensure "Training and Up-skilling" of rural folks for the use of toilets and water outlets, avoiding leakages of water. Up-keep sanitation. Local skills be developed to operate, maintain and repair electric appliances and water pumps.

6. Transport Sector

i)   This sector is one of the major culprit of causing steep pollutions. Stringent standards against emission output on auto-manufacturers has been in vogue in advanced countries. India must hasten to enforce equivalent measures (Bharat 6 to be advanced). Heavy carriers must be reined in; electric cars with advanced batteries and hybrid vehicles are a must.

ii) Public transport systems be upgraded by numbers and up-gradation of vehicles. Metros for major cities be hastened. Inter-city fast electric trains up to about two hundred kilometers are desirable.

iii) Long-haul goods must be transferred from diesel run carriers to goods trains.

iv) More powerful and longer lasting batteries are advised.

v)   Higher taxes on purchase of more than one car for private use.

vi) Develop second generation bio-fuels for blending with petrol/diesel. A word of CAUTION is necessary. Such of the bio-fuels produced using sugarcane and maize must not be at the expense of human consumption (foods and cereals). According to Jean Ziegler, UN's rapporteur warns, if such bio-fuel demand is not contained, the possibility of, hundreds of thousands of people deprived of their food, looms large. Instead, agricultural plants like JALOPHA, to manufacture diesel is a better option.

7. Agriculture

i)   Surprising though, the Agriculture sector is also a major contributor of harmful emissions. Chemical Fertilisers and pesticides give rise to methane and nitrous oxide. Methane can be converted in to methanol which can replace 'oil technology'.

ii) Photosynthesis is a natural process to maintain balanced ecological environments. Plants siphon off CO_2 and water vapour. A chemical process is available to trap CO_2 and convert it in to methanol. Methane produced on large scale, as already stated, can replace 'oil technology'.

iii) Organic farming is gaining roots. Organic manure costs Rs. 500 per acre against Rs. 2000 per acre of chemical Fertilisers. And, output is higher by 10% . Farmyards are already yielding organic manure. Besides, city garbage and sea algae are also sources of organic manure. The foods thus, produced are fetching better prices because they are more healthy. Tamil Nadu has made significant progress in organic farming, (readers can contact, for full particulars, Mr. M. Manimaran, at Kothavasal Village, Nanniam Taluka, Tiruvar District, Tamil Nadu (Reported in The Hindu, dated 13-9-2007).

Incidentally, sea algae avoid rats which harm rice crops.

8. Real Estate Construction Sector

Green Buildings and Construction Works for Residential And Commercial Purposes:

i) The Government Of India has brought out the "Energy Conservation Code 2007" - ECBC, for Green Buildings which is applicable to buildings of 10,000 sq. ft. or more, commercial buildings with a connected load of 500 kW or demand of 600 kVA. In it are norms for the energy efficient design and construction of buildings reduced energy requirement, environmentally benign design and construction techniques, and enhanced comforts for residents in it.

Occupancy sensors are proposed for residences, school class rooms, offices and any area of less than 300 sq. ft. The power will be switched off in non-occupied areas. Sensors of different kinds are suggested for control of electrical lighting vs. natural lighting for automatic switch-on or switch-off properties.

The scheme/Act if fully implemented, (overseen by local, state and central bodies), in the entire country, will save 1.7 billion units of electricity, worth Rs. 80,000 crores (2010 prices).

ii) Develop low cost climate-friendly dwellings for rural areas, providing for application of bio-gases, and solar devices for cooking, heating and cooling.

iii) The boom in building construction is a necessity for developing nations. Builders should not use potable water. They should not deplete water tables; conserve all water sources.

iv) Maximum use of efficient lighting, heating and air conditioning systems. Improved insulation, use of solar heating, recycling of fluorinated gases for air conditioning, automatic switching on/off lighting and air conditioning systems. Embedded solar cells in outer cladding and glass panes be considered.

9

EXAMPLES OF INNOVATIVE AND EXEMPLARY MEASURES OF MITIGATION AND ADAPTION

City Based Examples

i.   **Singapore** A city state is world renowned for its NEWATER Project, which Re-Cycles its sewage to obtain 20 million gallons of water to WHO standards. 6% of it is used for drinking too. The project uses dual membrane and UV technologies.

ii. **Vienna, Austria** The city has built 3 km long 30 meter deep Wastewater Tunnel which has an automatic pumping and distribution system that controls discharge of waste water to prevent pollution of river Wien; its water is used for drinking. (*Delhi Government may note*)

iii. **Portland, Oregon** There are a few streams girdling this town. A 2330 feet "bioswall" plus a number of check-dams have been built to stop 50% of silt and pollutants getting into river Willamete. This is termed as "Bio-drainage" system.

iv. **Tokyo, Japan** On top of a Sumo arena, rain water falling and collecting on a 8400 sq, meter area is harvested and drained into a 1000 cubic meter tank. The water is used for flushing and air conditioning. 750 buildings (possibly more) use RWH systems. The filtered water is used for drinking. In an emergency, the collected water is used for fighting fires.

v.   **Surat, India** Some years back Surat was devastated by 'plague'. The city municipality installed Micro-planned Solid Waste Management (SWM) systems in 52 wards to collect and dispose solid wastes; recycle and re-use. Hopefully, there will be no deathly epidemics again, Surat is now one of the cleanest cities in India, though at one time, it was one of the dirtiest.

vi. **Dwarka**, in Saurashtra is one of the driest areas in India. Yet, it has plenty of water in summers. Centuries old rain water harvesting practices in '*Tankas*' is its secret. 60% of the new houses have '*Tankas*' built in their premises. (on the other hand Baiji-ka-Talab in Jodhpur and the Mehrauli 'baoli' remain dry in rainy seasons, without RWH)

Urban Drainage Planning

In the recent past, excessive rains in Chennai, Srinagar, Uttrakhand have flooded vast areas, in fact, entire urban areas where collected waters have caused colossal losses to life and properties. Mumbai, Kolkata and Delhi suffer every year in monsoons with knee deep waters resulting in disruption of normal life. Besides poor drainage systems, 'heavy' population burden on metros, emboldened builders in connivance with politicians to encroach on open areas, to build apartments. The open areas were otherwise for absorption of excess waters. In fact, 40% of the mangroves along the Mithi River and Mahim Creak, have been illegally encroached on by Mumbai's builders. As a result, sea waters spill over into residential areas during high tides. Encroachments on Sunderban mangroves make conditions unbearable in Kolkata during rainy seasons and high tides. Delhi's builders have illegally used 60% of open areas for raising commercial malls plus offices and residential colonies. Poor maintenance and inadequate de-silting before monsoons, means our roads are waterlogged, and cannot be navigated.

Similar problems are faced annually in towns, world -wide (recently in Spain, the state of Louisiana-USA etc.).

The problems so caused underscore the importance of Urban Drainage Planning.

The best global models, plan integrated systems of public health and sanitation and water management with local hydrology. Broadly, the processes are planned to deplete the collected rainfall floods, in the fastest manner possible. Environmentally sound systems have adequate means of draining huge quantities of water, and collecting such waters for reusing. Of course, the solid wastes are segregated for disposal or recycling.

Suggested Measures for Avoidance of Water Logging in Monsoons

A report by the United Nations Environment Protection (UNEP) suggests the following measures to avoid water logging during monsoons/heavy rainfalls:

i)   A single authority to plan, construct and maintain drainage systems for a city.

ii) Separate drains for flood/storm water and sewage flows.

iii) No illegal encroachments along drains and water bodies. Inspect and remove encroachments before every monsoon.

iv) De-silt urban drains regularly, with vacuum or jetting pumps. The drain width and gradient must be generous. In the long run, piped drains require less maintenance, less costly and more reliable.

v)   Industrial effluents must be treated, filtered of solids, before letting them into rivers or seas.

vi) Rainwater Harvesting must be mandatory in metros and A class cities.

*Delhi can bridge the supply demand gap by 2021, by adopting RWH

vii) Porous pavements (non-muddy materials) for parking areas, and side-walks. These help in removing excess water.

Innovative Schemes to Avoid Water Logging

World cities are abandoning the conventional approaches of open drains, ditches and pumping, for example:

i) **Tokyo**, is digging infiltration pits to run offs sink into soil.

ii) **Lyon, France**, is building gravel filled filter drains to prevent sediments reaching recycling pumps.

iii) **Ontario, Canada**, is constructing detention ponds near rivers to reduce the down-stream flooding.

iv) **Zurich, Switzerland** has adopted "stream day-lighting"; 16000 meters of the city's streams have been revitalised and 31% of the extra rainwater is diverted into them to prevent flooding the city. The city is installing rot-proof, impact resistant polyester silt traps which capture increasing quantities of silt from surface runoffs to keep drains clean

v) **MCD Delhi** Garbage landfill sites start rotting within five years and remain operative for 15-20 years. The rotting garbage gives off Methane and CO_2, which if allowed to escape freely, add to global warming. Delhi has 4 large landfill sites, which can produce large quantities of GHG. Either these gases can feed houses through piped systems or earn "Carbon Credits", by destroying them, if the quantities are not sustainable. Japan has granted $ 48900 to MCD for a feasibility study to check their concentration and quantum over the years. Developed countries buy CC @ $4 to $ 10 per ton of gas destroyed. As per estimates, the 3 major landfills at Ghazipur, Okhla and Bhalaswa are capable of generating 750 tons of gasses daily and thus earn Rs. 400 crores annually. A tidy sum for the poor MCD!

Incidentally MCD did set up a piped-gas distribution system fed from the 4th landfill at Timarpur near Balak Ram Hospital. A power station can also be set up if a network can be arranged of the 4 sites in Delhi.

(H.T. July 16, 2007. Articles by Vipul Mudgal, Saikat Neogi, Cooshalle Samuel, Renuka Bisht, and Vipul Mudgal UNEP et al, World Bank, ADB, Planning Commission, IWMI, CSE) l

Note: But the scheme was not pursued seriously or sincerely

vi) **Bangalore** Water Supply and Sewage Board is using BISON (Bangalore Info-systems on Network) to manage and maintain the city's water and 'waste-water' network.

vii) **Kolkata** is using a 3D flood simulation model to maintain its various flood situations. The city has arrested *flow of sewage* into Hoogly river, by building sewage treatment plants to underground sewers. **Hyderabad** similarly has laid such systems in industrial areas and thus, reduced inflow of toxic effluents into Hussain Sagar Lake. Besides enhancing capacity of such water resources to withstand monsoon overflows, they augment percolation to groundwater tables.

viii) **Calicut** has adopted Mission for Application of Technology to Urban Renewal and Engineering, MATURE, to undertake an integrated development plan and to regenerate the city's ponds and model its drainage wastewater facilities.

ix) The UN Environment Programme's Integrated System, as administered in **Kerala and West Bengal** has shown encouraging results. Biochemical reactions use Sun's rays for recycling waste water. Thus, treated wastewater is led into wetlands for developing aquaculture and irrigation. Additional employment is a bonus.

However, it may be recalled, many years back, the Government of India sanctioned a grant of Rs. 1200

Steps Towards A Green World

crores for **Mumbai** to improve its drainage systems. UNESCO's Institute of Water Management reported that the programme could not succeed without improving governance at municipal level.

x) **Central Road Research Institute** recommends 'echo-phalt' road technology which uses Drain Asphalt Modified Additive that sucks surface water from roads to keep them dry. A trial road was laid in 2006 along Dhaula Kuan on NH8.

xi) No city, in India has overcome its woes of heavy rains, cleaning City Slums and Unauthorised Colonies. There are no specific examples (to the author's knowledge). Perhaps, a likely solution is to provide their residents with decent living abodes, at modest costs but in clean and airy areas. These sectors must have drainage and flood-water removal systems as in the better parts of the metros and big cities.

Cities' Managements increasingly realise that good drainage systems are the foundation of sustained development. They will have to take cognizance of "freak" weather and exceptional rainfalls when least expected. The principal life support systems, electricity and water supplies, also breakdown; spreading miseries all round. Judith A. Rees, in a paper in 2006, read at the Global Water Partnership illustrated that the planned use of urban waste waters can provide reliable water supply for agricultural communities . Such peri-urban irrigation can yield cheaper vegetables to city dwellers. Polluted waters of rivers, if properly treated, can become assets instead of a nuisance/liability. The returns can be, expenditure to income, in the ratio of 1 to 5 according to a 2004 WHO study.

Interesting and Useful Examples of Individual-Level Efforts to Preserve Environment

One need not be an activist writing pamphlets for an NGO or brandishing placards nor there is need to shout hoarse against Americans, Chinese and Australians who pollute our planet. Ordinary humans, like you and me, are by myriad examples, changing damaging life styles. They have given up air conditioners, use cycles instead of cars (which they can afford) or just refuse to use plastic bags. A few among millions they, by examples to be emulated, are adopting life-styles which are not easy.

i) **Shilpa Jain of Udaipur** gave up her easy life in Washington, D.C., USA where she owned Toyota Cammary car, apartment with central heating and cooling, perfumed toiletries and gym, gadgets etc. to a simple but useful life in Udaipur. She rides her cycle all over the town for her shopping and her other engagements. Compost is made from her domestic wastes, employs mud coverings to keep a shiny skin, etc. She advocates rethink about education, health and development policies through her Shikshantar movement in Udaipur and around.

ii) **Mukul C. Mahant of Guwahti** abhors brick kilns. They convert crop-giving fields to satiate builders' greed besides leaving large craters in ground. For his two story house, he used stones and hollow prefabricated concrete slabs; the house is totally brick-less, and Mr. Mahant does not require air conditioning nor even fans. About 30 of his friends have followed suit.

Although Mahant is an electronics engineering graduate from IIT Kgharagpur, he diligently studied construction processes and materials best suited to local climate and environment.

iii) **Mansatas of Kolkata** are even more adventurous than Mr. Mahant. The young family, living in a cluttered area of Kolkata, decided to build a bamboo house above their book-shop. Bamboo abode costs a pittance to the comparable pucca houses. Zui their 13 year old daughter, living in a tree-house is the envy of her friends. Though,

     Steps Towards A Green World

every monsoon, the roofs covered with gunny coverings leak, the young family weathers it happily. And they lived more than seven years in that anachronistic shelter in the heart of a concrete jungle. (Note this example may help very poor families and less rainy parts of India)

iv) **Indrakumar of Chennai** is however less audacious than his Kolkata counterpart. He has 150 potted plants on his house-terrace. Outer walls carry creepers (trained). Neither the roof nor the walls get hot. Indrakumar laughs, "I get free oxygen and air conditioning". Rain water is harvested only after passing over medicinal plants. Even the nearby tanneries' effluents do not dilute the fragrance and sweetness of the water. All bio-wastes are treated with Actizen, an enzyme to obtain liquid urea for his plants. Can anyone deny ingenuity of over 60 year old Indrakumar!

v) **Sukriti Chopra** fines everyone one rupee even her parents, if they bring home shopping and gifts in polythene bags. Prateek only 17 (when reported) made his parents sell their AC and Maruti car and allows no one to bring home polythene bags. The boy is studying the impact of sulphur and nitrous oxide content in the air. Ranmal Singh commutes on a cycle in Delhi. Ajai refuses to attend weddings in evenings. Young couples in Mumbai are getting wedded in pre-noon functions. All are saving waste of electricity. Uttamchand Shah got his son married in a jute and bamboo pandal and used castor oil lamps in Mumbai. He served aam-ras in earthen pots. His son khadi and the bride's sari was of silk spun without killing silk-worms.

Note: Some of the examples given above may not be replicable on large scale. But many others can be advantageously adopted. There are many young people who are eager to prevent environment degradation by some innovative ideas.

10

SOLAR, WIND AND OTHER RENEWABLE RESOURCES OF ELECTRIC POWER

A. Solar Electric Power Generation

Solar Radiations give out 'Light, and Heat"; the two primary energies for human uses for centuries at end. Electricity, the secondary form of energy, came into use, primarily, in 20th century. Almost all nations are concentrating on installing solar power stations to cut down GHG emissions. The International Energy Agency, predicted, in 2014 that Solar Power will rise by 2050, to 27 % of the worldwide electricity consumption; most of it in China and India. Electricity from Solar Light is of (i) Photovoltaic Type (PV), (ii) Concentrated Solar Power (CSP) from heat and (iii) Hybrid Types- the combination of these two and with many other forms. All these are explained below:-

1. Photovoltaic Electricity Panels and Plants

In 1880, Charles Fritts obtained electric current directly from light. Fifty years later in 1931, Bruno Lange, a German, developed a solar cell with silver selenide. This selinium cell converted hardly 1% of light into electricity. Only in 1954, Calvin Fuller and Darlyl Chapin developed the silicon cellar of 4.5% -6% efficiency "pv" cell, which generated D.C.. The cost was as high as $286/watt, and it was initially used in mobile phones. However in due course, multiple cells were connected to form 'modules' which were connected to form 'arrays/panels' to achieve the desired voltage;

the DC voltage alongside is converted into AC, of desired frequency/phase, using inverters.

Solar Lanterns became handy for dwellings in villages. Far flung areas, with continuous sunlight over the year could have benefits of electricity, avoiding use of long transmission lines connected to a 'grid/base' power station. Next stage was of a large panels mounted on roof tops to meet needs, partially, of a dwelling or an office. It became necessary to connect such installations to electric grids, to make judicious use of solar power and utilise the grid power, only, as and when necessary. And since 2000 onwards, high capacity power stations have been installed with sufficiently large numbers of solar power panels. Such plants are mainly situated in arid deserts where sunshine stays high for most of the year. Solar Power can be stored, in batteries too, besides, connecting to the country's grid. And thus, solar power is of universal benefits.

China has the highest installed capacity, PV type, at 43000 MW. However, the largest PV power Plant, 579 MW capacity, is in California, USA. In this country the total capacity is believed 3,520 MW, almost 90% in California. India, is on fast pace in this field at about 5GW presently, but the plans are to leap-jump to about 20GW by 2018 and 100 GW by 2022; highly ambitious intentions. Rajasthan, Gujarat, Maharashtra, Tamil Nadu and M.P. are in the forefront.

Note: The data given above and those below cannot be vouchsafed; All countries are racing fast and up-to-date figures may vary. The intention is to inform you the importance of solar power, PV+ CSP+ Hybrid types, in the coming years.

Israel has made an ingenious use of PV panels, which are mounted atop poles in agricultural fields. Besides, gaining electric power, there is no effect on the agriculture. Similar

Photovoltaics

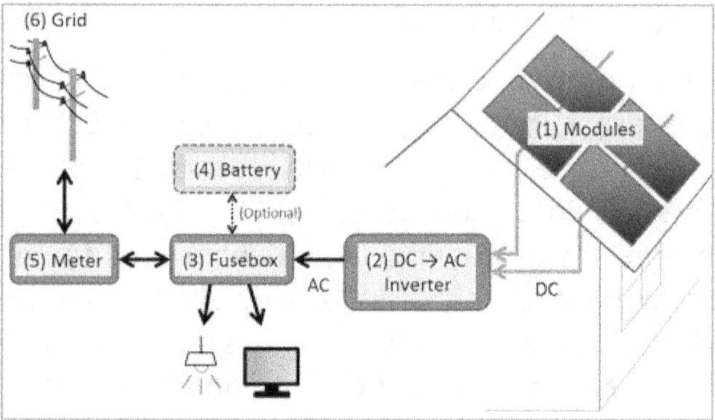

Fig. 10.1, Schematics of a grid-connected residential PV power system

high pole mounted panels can sit on residential buildings in a congested area.

2. Concentrating Solar Power (CSP) Stations

For more than 50 years, the heat energy from the Sun, has been used for water heating at hotels, offices and residences. All wash rooms and showers for bathing used the water so heated. CSP plants were first installed in 1980s. A wide range of concentrating technologies, e.g. parabolic troughs, linear Fresnel reflectors, Stirling dishes and solar power towers. These equipments track the sun and focus sun's radiations, heating a fluid by the concentrated sun's heat. The bolstered temperature of the fluid, raises steam which in turn utilises the conventional turbo-generators; a great advantage of CSP.

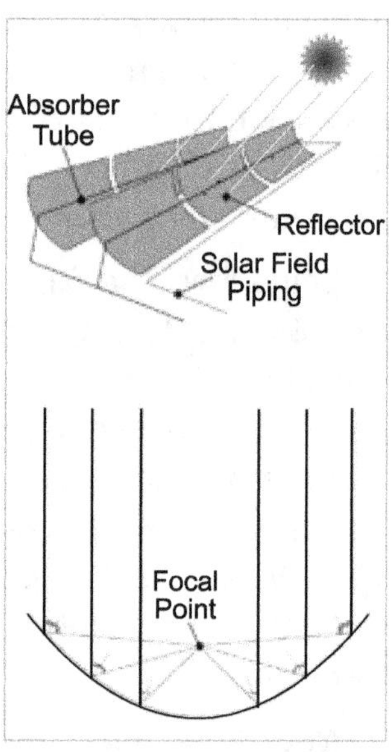

Fig. 10.2, Concentrated solar power. A parabolic collector concentrates sunlight onto a tube in its focal point

A Parabolic Trough consists of a parabolic reflector which focuses the sunshine on a receiver-tube full of a fluid, above the middle of parabolic mirror. This type uses lesser space, per unit of power, than the other systems Parabolic Troughs are used at the USA's CPS, in California and Nevada, 377 MW station, the largest in the world, is located southwest of Las Vegas; Ivanpah CSP.

Compact Linear Fresnel Reflectors use many thin mirror strips, instead parabolic mirrors. The concentrated sunshine heats a fluid in two tubes. Linear mirrors are cheaper than the parabolic ones and more of them can be packed in the available space.

Stirling Dish type combines a parabolic concentrating dish with a Stirling engine. This version is high in efficiency and lasts longer. Australia has used a Big Dish of 50kW capacity in Canberra.

A Solar Power Tower has a number of reflectors (heliostats) to concentrate sun's rays on a central receiver atop a tower. Tower type is less costly and offers higher efficiency.

A total capacity of CSP plants, 2,541 MW is in the USA, followed by Spain, 650 MW.

India has just started. The World Renewal Trust of Brahma Kumaris, supported by GOI's Ministry of Environment and Forests is setting up a station for 20,000 residents at Mount Abu, Rajasthan. With German collaboration, the project is 'India One Solar Project. The station will have 770 parabolic dishes of 60 sq.m, each covered with 800 pieces of solar grade mirrors, capable of generating 1200°C temperature. The project cost is Rs. 80 crores, is 1MW plant in solar energy terms, but can generate 3.5 MW in 24 hours. The cost will come down when produced at industrial scale.

3. Hybrid Systems

In the future, CPV cum CSP combinations are quite likely to provide 24 hour electricity. PV and CSP stations can combine with other forms e.g. wind, bio-gas or diesel, coal etc., to modulate power output according to demand. Presently, there is no Hybrid installation in sight; except roof top solar generation in tandem with grid power.

4. Solar Electricity Cost

For well over two/three decades the per unit cost of electricity was very high; the least running around Rs.2o per unit. However, in the recent past this cost has dwindled to around Rs.4.5 only, quite comparable with other forms. In view of this, the Nuclear Power is getting relegated (see Nuclear Power).

5. Solar Power in Building Sector for Glass Panes

Buildings, both residential and commercial, are going to be very large users of solar energy. Intensive research is going on to develop film-type materials which can be pasted on window-panes and glass doors.

B. Wind Power

China, Denmark, Ireland, Spain, Portugal are the big holders of wind farms. The wind power installed capacity is growing at a fast pace; at 16.2% in 2014-15, from 318.6 GW to 369.55 GW (China 148 GW). The installations are both on-shore and off-shore where high velocity winds run for most of the year. In Denmark wind power is about 39% of its total installed capacity (Denmark has hardly any fossil fuels based generation.)

India had wind based installed capacity of 23.439 GW by March 2015.; Of it Kutch in Gujarat has 11 GW capacity. The predicted capacity by 2022 is 60 GW. The country has low wind tides and requires special designed wind-turbines, thus, mean Rs.6 to 7 crore per MW. Tamil Nadu,

Maharashtra, Gujarat, Rajasthan and Karnatka are the five top wind power producers.

Basically, wind turbines run induction generators, their output is connected to the grid or stored in batteries. The power in the latter format is utilised on 'Demand' needs.

C. Nuclear Power

Nuclear fission has been the well established resource for power generation. The heat released by nuclear fission is utilised to raise steam which then runs turbo-generators (at 1500 rpm only). The fission rods of uranium are shrouded and a number of them are fitted into steel vessels. These modules are then placed in blast-proof brick, cement enclosures. The fission rods have a designated life for generation of electricity but continue to throw out harmful radiations. Therefore their disposal is an onerous task; a liability for the nuclear electric generation. Nuclear stations require long times, up to 10 years for building and gestation requirements. The costs of fission rods and buildings, compared to other forms, are very high, although the running cost of generation is moderate. Recent accidents in Russia and Japan, resulting in big loss of lives and colossal damages to property besides lethal radiations spawning out, have had adverse effect on the growth of Nuclear Power. Some countries, Germany and USA are even dismantling their nuclear power installations. Of course, the headache of nuclear waste disposal is another factor.

India was one of the earliest countries to install nuclear power stations for civil purposes. All equipment, and technology were imported. Our research institutions failed to develop thorium based power production; we had to rely on imported nuclear power equipment, materials and technologies. The imports got banned when India tested its first nuclear explosions, for atom bombs, in 1978 or so. Another underground nuclear testing, made the world more closed for us. Dr. Manmohan Singh's, Former Indian Prime Minister, strenuous efforts succeeded in winning over America and the myriad bans were revoked. MOUs have been signed with Russia, France, and USA.. But, there is little progress on the ground; only one reactor has been installed in South India; the second unit may be installed in a few years.

439 installations in 31 countries, have a capacity of 376.8 GW, China's contribution being 148GW. Russia, South Korea, China and India are actively pursuing nuclear resource. But more than 100 nuclear installations are likely to be phased out. America, Germany, and many other European countries are closing down nuclear stations, subsequent to the disastrous accidents in Japan recently and in Russia earlier on .On the other hand Belgium, France, Hungary and Slovakia depend primarily on nuclear power. The future of nuclear power is somewhat uncertain.

D. Hydro Power Stations

Water has been a source of energy even before the Industrial Revolution; the village folk using falling water for flour-mills. Soon after the invention of electricity, French and British developed turbines to make use of the dynamic energy of falling waters from a height. It was mega size hydroelectric stations came into being. Hoover Dam in USA, and dams on Ontario River, Canada, were the fore-runners in this genre. Besides being a renewable source, the water released from dams got used for agricultural purposes, too; many cities received their drinking water needs also.

Hydro power generation at high rise dams are mega power stations, so are some medium height dams, too. Then there are mini, micro, and pica dams generating a 100 MW, 100 kW, and 5kW respectively. Besides the dams types, there are run of the river hydro-stations; at times waters are run through tunnels to make use of height-fall on the other side of the mountainous terrain. Tidal-wave generation of electricity, at sea shores, is also possible.

China, once again is the largest producer of hydro-electric power. 150 countries generated 3,427 terawatt hours of hydro electricity, in 2010, of which 32% was contributed by Asia Pacific region. Presently this type forms 16 % of global electricity. A growth rate of 3.1% annually is visualised. Hydro-electricity about the cheapest, only 3-5 cents per unit, at stations 10MW and above capacity. Another advantage that power generation can be easily varied at a fast rate. Hydro Stations, therefore, used to meet peak demand. There is hardly any GHG releases, a major merit for protecting environment.

There are a few major drawbacks of high dam hydro power stations. Firstly huge amounts of silt accumulates at those sites requiring frequent de-silting. Secondly, God forbid, in times of heavy rains if the dam-bursts, all the populations, situated on their down-flow canals routes, get inundated underneath fast flowing waters. Thirdly, vast populations, wild life and agriculturalists are displaced, on account of vast areas acquired for the high-rise/ large dams.

———

11

FORTHCOMING TECHNOLOGIES
(SOME SUGGESTED FOR INDIA)

In earlier chapters, measures for mitigation of GHGs were described along with a number of solutions in existence. But, those measures can/will only partially meet India's needs of the future i.e. (a) extend electricity, 24 hours, to all villages and raise, at least, to living standards prevalent in major cities in the country, (b) build up electrical (other forms of energies, too) resources to meet rising populations and (c) rise to the living standards of middle class populations, in developed countries. The simple answer is "usher in new technologies".

The onslaught of solar and other forms of renewable-sources of electricity have been described elsewhere. Given below are " innovations" which can be harnessed with persistent researches on our own.

A. Carbon Trapping and Sequestrating of CO_2 from Air

i) Energy companies have been attempting to trap the CO_2 emissions of their utilities and collect them in vessels and transport them to geographically remote and dispersed locations. The objective being to pump the collected CO_2 in to bowels of the earth. But the collector-emitters-vessels cannot be large to be carried on trucks, but the smaller vessels are economically unviable. Hence a question mark hangs on.

ii) Photosynthesis, the natural way of recovering CO_2 from air. Chemical reactions, instead, can be hundreds

    Steps Towards A Green World

of times faster than photosynthesis and less costly too. Prof. Klaus Lackner of geophysics at Columbia University, USA, saw his young daughter, passing air in to a solution of sodium hydroxide. Oxygen and sodium carbonate were released in to the atmosphere. He realised that his job was limited to isolating sodium and CO_2 which can be stored.

The professor co-founded Global Research Technologies (GRT) to evolve a new methodology for isolating sodium and CO_2 and storing the latter. In April/May 2007, GRT researchers successfully demonstrated that a 'special material-proprietary' extracted 10 times CO_2 from the air. According to the process, the normal air is sucked inside a tower where it is separated in to CO_2 and O_2. While residues of air and O_2 are released in to atmosphere, the trapped CO_2 goes in to storage in the tower. According to GRT (a) the tower trapped CO_2 is 100 times faster than a tree of the same size, and (b) the cost of removal of CO_2 is 30 USD per ton (2007 prices which, according to GRT can be set off against raising the price of petrol by 10%). To maintain CO_2 concentration "at the pre industrial era", it is estimated that 1,100 billion tons of CO_2 have to be removed from the atmosphere.

We may see in 15 years, if the proposition is adopted by states or energy producers, clusters of towers, like wind mills, dotting fields and highways.

iii) David Keith, University of Calgary, Canada, is also working on separating CO_2 from air. He demonstrated his process at a pilot plant a few years back. Probably the next step is to convert CO_2 gas in to solid state. When magnesium silicate hydroxide reacts with CO_2, it forms magnesium carbonate, a solid. This is happening naturally at a very slow pace. Since, India owns huge deposits of the said material, on the Deccan plateau. Some research towards this technology is called for, by our researchers.

Note: Research in this area, in India, is showing good results.

iv) In India, at the Tata Institute of Fundamental Research(TIFR), a team under Mr. Vivek Polshettiwar, is designing a fibrous nanosilica, KCC-1 (a nanomaterial amine) for binding with CO_2. The fibrous material enables higher efficiency of capturing CO_2.

More details at: www.nanocat.co.in (Amitabh Sinha, Indian Express July 24, 2016)

B. Some Smart Innovations, Ready To Be Adopted, Particularly in India

i) Smart Grid of Electricity. Presently, in Indian Transmission Grids of electricity, the losses are too high, inadmissible. India's annual 'power distribution' losses amount to $50 billion. "Smart Grids", control from remote locations, generation and flow of electricity to areas where it is needed. Smart meters transmit data regularly, of the consumption of power as well as of the health of the 'Grid and efficient routing from generation to distribution points". India will need such systems, to integrate efficiently, the thermal power generation with those from solar and wind power locations (which are often of irregular kind).

The President, India Smart Grid, claims that "the extra high voltage that Power Grid Corporation and POSCO is managing a very intelligent grid comparable to the best transmission systems in the world. On the distribution side, some private discoms are experimenting with advanced solutions such as smart metering, demand response and energy storage".

ii) Change over to LED lighting and efficient air conditioner systems. Use of solar power on commercial and residential buildings including affixing solar power films on window glasses. Highly efficient air conditioners, and advanced cooling systems are coming into markets.

iii) Bio-Diesel (Gen-Next Fuel). In 1893, Rudolf Diesel, tested vegetable oil as fuel. He subsequently 'diesel' for autos. Petroleum based fuels, for various types of engines, the bio-diesel got ignored till the oil crisis in 1973. Scientists, many in India too, have succeeded in making each unit of bio-diesel, gain 3.2 units of energy; an exceptionally efficient alternative fuel. In August 2015, the GOI ordered that 5% bio-diesel +95% customary diesel, termed B5 Blend, be sold in New Delhi, Vishakhapatnam, Haldia, and Vijayawada. Railways and Public Transport Corporations are to adopt B100, pure bio-diesel.

Bio-diesel is considered 'biodegradable', under ideal conditions and is "non-toxic" too. A study by the US Department of Energy, that the production and use of bio-diesel releases 78.5% less CO_2 than the conventional diesel. This diesel generates only 50% of ozone-forming hydrocarbon emissions.

Bio-diesel cost, of production, may range between Rs.35 to Rs. 45 per liter. Using advanced catalytic systems, and some other advanced technologies, can bring down its cost. But the bigger challenge is to achieve commercially viable capital costs.

iv) Electric and hybrid cars have been in public domain for long. But even the latest electric cars are much too costly for common adoption of such vehicles. The challenging tasks are affordable high capacity vehicles and installing 'charging' facilities at each and every, present day, petrol disbursing outlet. Hybrid kits, to fit onto, petroleum vehicles, must be affordable, requiring no maintenance and be entirely safe. In spite of all hurdles, one can hope to see their universal adoption in a decade.

v) Improved Supercritical Type Thermal Stations: Some technological breakthroughs, by researchers in Denmark, Germany and Japan, have helped increase efficiency, 30-40% gain in energy, of "Super Critical Thermal Power Stations"; lesser cost as well as lesser CO_2 emissions. About 60 "Ultra Supercritical Units, were in operation in 2015. India too, is setting up such a power station, in which BHEL, NTPC and Indira Gandhi Centre for Atomic Research are co-operating. A 4000 MW power station, using such technology, can save 4 million tons of coal in a year. And the cost per unit is likely between ₹ 3 to ₹ 3.25.

It may be mentioned, once more, that there is a wide ranging campaign to root out coal based thermal power production.

Note: On April 5, 2016, 3300 MW Of SASAN ULTRA MEGA POWER PROJECT, of a 3960 MW Project, the largest in the world was commissioned in Madhya Pradesh by Reliance Power. It is with Chinese Colaboration. The cost per unit is Rs. 1.19, the lowest in the world.

vi) Organic Rankine Cycle Technology - Waste Heat Converters. USA and Israel are using ORC technology to produce electricity from waste heat released from boilers and turbines at thermal power plants, furnaces and boilers common at many industries; it is like using converting garbage into bio-fuel; though the efficiency conversion is low. Thermax, a Pune based corporation is presently developing a 100 KW unit; hopes to develop commercial size units in five years, with a view to reduce present day cost of 11-12 crores rupees per MW to around ₹ 7 crores per MW. The cost of imported technology can be 30% costlier. A major advantage of

ORC plants is rather easy integration within the same industry. Awaited anxiously!

vii) An extraordinary source of power can be obtained by running wheels of a cycle. Manoj Bhargawa, a billionaire, developed a power-generating cycle by using 70% parts of a common cycle and adding others by his ingenuity. One hours cycling generates sufficient electric power to use a 25LED bulb and charge a cell phone and a tablet. Who for, is a natural inquiry?

At least 32,227 villages, in India, over 900,000 households have no electricity. One hybrid-cycle costs mere Rs. 12000. Bhagava wants production of 100 million such cycles. Unsophisticated though, here is an innovation that speaks for the poor. Will the governments care?

viii) Large areas are left behind as waste lands after mining coal or other minerals. Central Coalfields India Ltd. has grown large plantations on reclaimed land of Piparwar mines. SECL has developed beautiful flower parks on reclaimed land.

(Excerpts from India Today December 14, 2015)

The developments and innovative ideas of replanting reclaimed lands require serious thinking by GOI and large and small businesses. Solar Power is not the sole answer!

C. Radical Methods to Generate Energy

i) Scientists at Basel, Switzerland, stumbled on "Hot Rock Energy" accidently in 2006. While drilling deep in to the earth's crust, (well away from 99% of the earth's interior) where temperatures vary from 500 °C to 700 °C, to tap its energy and thus break new ground, literally. A shaft was bored by a 58 meter high drilling rig, not far from apartments. Water was poured into the well; the temperature of the water went up to 200 °C. Apartments near the site of drilling felt tremors, and their walls as well as tiles in bath rooms got cracked (Hindustan Times). The prediction of obtaining electricity from 'Hot Rocks', is very encouraging. Besides, the inexhaustible source of the geothermal energy, it is clean and has zero impact on climate. Basel scientists believe that this source can provide 2,50,000 times this world's needs. Though authorities have stopped Geo-Power Board to pursue its investigations, any further, the Corporation has not given up.

ii) Radical (may be bizarre) Options.

    a) For reducing solar heat, paint deserts with white paint.

    b) Cocooning the earth in tin foil. Astronomer Roger Angel, Arizona University, proposes a giant 20 m tons sunshade between the earth and sun. Only 1.8% of the sunlight can be thus blocked. But, the cost will be a few trillion USD, a prohibitive proposition.

    c) John Latham, National Center for Atmospheric Research, and Stephen Salter suggest creating a fleet of 50 unmanned ships blasting seawater into the air to form highly reflective clouds which would provide substantial cooling effort. Each vessel will cost a few million US dollars. And spraying about 10 kg. water per second could cancel out a year's CO_2 emissions. And for how long? No, ready estimates.

iii) Nano-technology could be the saving grace. Research can put out the silver bullet technology to replace conventional fuels within 20 years, according to inventor Ray Korowol. At MIT, researchers are trying to make "super capacitors" which can be recharged at least hundred more times than ordinary batteries.

Note: The author is hopeful of the Nano-technology.

iv) Converting energy of sea waves in to electricity, surfaced initially in 1974. Generally, the wind turbines were modified to use marine power; the methods were complicated. Oscilla Power, based in Seattle-USA-employs "magnetostriction", a phenomenon in which ferromagnetic materials (which get magnetised easily, like iron) change their shape slightly in the presence of a magnetic field. This process works in reverse too, i.e., a slight change in shape alters magnetic character of a material. The Oscilla process employs bars of iron and aluminium (which are highly ferromagnetic) which are compressed one part in 10,000 to have the desired effect. The huge force required to compress the ferromagnetic bars, is provided by the oceanic waves. The change in the magnetic field is utilised to generate electricity. It means that the electric generator has no moving parts. The oscillating generators consist of two large objects connected by cables. At one end of these cables, floating on the sea surface, is a floating buoy that contains alloy bars, magnets and coils together with sets of hydraulic rams, which can squeeze the bars as desired. At the cables', other ends hangs a 'heave plate', which is kept stationary by a combination of inertia and drag of surrounding water. While the buoy rises and falls with the sea waves, the 'heave plate' stays more or less stay put. The tension on the cables increases and decreases. The changing tension drives the rams. The whole system is connected to a second set of cables that moor it to the seabed.

Mr. Rahul Shendure, Chief boss of Oscilla, claims that a fully developed device, designed to deliver 600 kilowatts, will be ready by 2018. The machine will be costly, but it will operate for decades, because all its parts being stationary. The cost of producing electricity from the Oscilla generator will be around 10 cents kwh. In due course the oscillating generator should achieve cheaper than 10 cents per unit electricity. Wind farms off shore produce a kwh at 16 cents, though the on shore wind mills produce a unit at 6 cents.

(Note: A report in The Economic Newspaper Limited 2015; Indian Express 11-11-2015)

(v) Solar Energy A New Approach—An Artificial Leaf.

The researchers at the University of California Berkeley are researching on storing solar energy for long periods on 'artificial leafs'. The first component, called 'catalyst' has long filaments made of nickel sulphide or platinum called nano-wires. These nano-wires, first turn the sunlight in to electrons. Then the filaments split water into hydrogen and oxygen. In the second part of this artificial leaf, a genetically modified version of naturally occurring methane producing microorganism "methanosarchinabarkeri" is introduced. This microorganism takes up the hydrogen, obtained from splitting water, and combines it with carbon dioxide to make methane- the biggest component of natural gas that is used as fuel.

The new fuel generated from synthetic photosynthesis will absorb carbon dioxide from the air and convert it into liquid fuel. This process stops release of GHG into air.

The research team, wants to use this photosynthesis on "atomic and molecular scale" to develop more sustainable resources.*{Indian Express 13-08-2015}*

Converting CO_2 into Stone

Dr. Juerg Matter and his research team, at Southampton University, in Britain, are pursuing a project "CarbFix", to convert CO_2 in to rocks of calcite and magnesite- stable rocks. The team worked between January- March 2012 at Hellisheidi geothermal power station, near Reykjavik, Iceland. Geothermal energy uses hot groundwater to drive steam turbines. H_2S and CO_2 are released. H_2S is a noxious pollutant, which is scrubbed and stopped from reaching the surface water of the sea. Only CO_2 collects on the surface.

The team collected 175 tons of carbon dioxide, mixed it with a mildly radioactive tracker chemical, dissolved the mixture in water and pumped it into a layer of basalt half a kilometer below the surface of sea water. 95% of the injected CO_2 was mineralised in less than 2 years. (Note. Basalt is full of elements which react readily with CO_2. Over millions of years CO_2 is removed from air due to weathering. This research aims to speed up the natural process). The problem of absorbing H_2S has also been, to a great extent, solved. The iron in basalt converted H_2S into pyrites.

A large scale test, burying 10,000 tons of CO_2 and 7.300 tons of H_2S are being converted presently at Reykjavik Energy. The result of this experiment is awaited. Further, the fossil-fuel based thermal power stations have SO_2 as the additional pollutant; it is different from Hydrogen Sulphide. Therefore, scrubbing may still be necessary. Basalt is predominantly available on the ocean floor. Basaltic regions are available on dry land, but may not be conveniently located near thermal plant sites.

In spite of doubts and constraints, Dr. Matter's proof of the principle of chemical sequestration of CO_2 in rocks is worth pursuing.

(The Economist / Indian Express June 13, 2016. Excerpt of a Paper from "Science").

12

EFFORTS TO COMBAT CLIMATE CHANGE

Updated December 11, 2011 10:49 AM NPR News Staff

While nations wrangle over a new global treaty on climate change, the question on many minds is: What happens next?

Even though the key portions of the Kyoto Protocol was set to expire by the end of 2012, many of the world's major green house gas emitters started to set national targets to reduce emissions. They started and continue to forge their own initiatives to meet the goals.

Some are focusing on curbing deforestation and boosting renewable energy sources. Several nations are experimenting with cap-and-trade plans: Regulators set mandatory limits on industrial emissions, but companies that exceed those "caps" can buy permits to emit from companies that have allowances to spare. In some cases, it's not clear what countries are doing to meet their stated climate goals. What is clear is that and they fall far short of what would be required to stabilise the planet's atmosphere.

Here's a look at what nations are doing:

Parties, Nations

As of 2014, the UNFCCC has 196 parties including all United Nations member states, as well as Niue, Cook Islands and the European Union. In addition, the Holy See and Palestine are observer states.

General

1. Peak GHG Emissions in tons per capita by 2030: Agreed by USA- 23 and China- 14, India offered 10 at Lima. India can commit not to exceed China's present day emissions till 2040. The Lima COP can ask USA and China to commit better reductions than the two countries have agreed to at Beijing recently.

2. USA at a recent G20 meeting in Brisbane, Australia, agreed with China to reduce its GHG emissions by 26-28 per cent over its 2005 levels. China promised to its peaked emissions latest by 2030 and 20 % of its energy will be from non-fossil sources. China's emission per capita by 2030 will be 13.25 tons; population 1.39 billions. EU has already committed to 40% reduction by 2030.

Country-wise Efforts

Australia

Australia has set a national goal of reducing greenhouse gas emissions by **5 percent below 2000 levels by 2020**.

Australia didn't sign the Kyoto Protocol until 2007, after its Labor Party took control of government, reversing the previous administration's policy. Under the climate pact, Australia agreed to hold the growth in its greenhouse gas emissions to 8 percent above 1990 levels for the 2008-2012 period. By and large, Australia has met those targets, mostly by reducing deforestation and land clearing.

In November 2011, Australian lawmakers approved an ambitious carbon trading plan — the world's largest outside of Europe. Under the plan, Australia's 500 worst polluters would be forced to pay a tax on every ton of carbon they emit starting in July 2012. By 2015, the nation plans to move to a full-on, market-based carbon trading system. Australia says it plans to link its carbon market to one set up in neighboring New Zealand. That might make it harder to dismantle the market if conservatives win back control of Australia's government in 2013.

Brazil

Brazil is aiming to reduce its emissions to 1994 levels and cut deforestation by 80 percent from historic highs by 2020.

Brazil's National Climate Change Plan is focused on expanding renewable electric energy sources and beefing up the use of biofuels in the transportation industry. The country is also focusing heavily on reducing deforestation rates: It's hoping to eliminate illegal deforestation and bring the net loss of forest coverage to zero by 2015.

But, a proposal to loosen Brazil's deforestation rules is currently making its way through the legislature. If enacted, critics say the changes could create more opportunities for logging.

Canada

When Canada signed onto the Kyoto Protocol, it committed to reducing its greenhouse gas emissions by 6 percent below 1990 levels. It later proposed a new, less ambitious goal to reduce emissions by **17 percent from 2005 levels by 2020**, a pledge that matches the U.S.

Canada did little to try to meet its obligations under the Kyoto Protocol. Indeed, today, the country's emissions are 17 percent above 1990 levels — in large part because of emissions tied to the dirty business of extracting oil from Alberta's tar sands.

According to a Canadian government, report released in mid-2011, emissions from tar sands will more than cancel out the progress that Canada has made in shifting its electricity generation from coal to natural gas. By 2020, the report projects that Canada will fall well short of its stated emission-reduction targets.

China

China is the world's biggest producer and consumer of coal — and the No. 1 emitter of greenhouse gases and the second-largest consumer of energy. But it's also a developing nation — which means that, like other developing nations, it isn't required to lower its emissions under the Kyoto Protocol.

Still, China's coal resources aren't infinite, and as the country finds itself importing more of the fossil fuel to power its growth, it is also aggressively pursuing renewable energy sources. Chinese leaders have said they want non-fossil fuels to account for 15 percent of the nation's energy sources by 2020. Under a law passed in 2005, Chinese power grid companies are required to purchase a certain percentage of their total power supply from renewable energy sources. And China provides extensive subsidies to its clean energy sector — like the U.S., it hopes that green tech jobs can fuel future growth. Even so, many analysts warn that weaning China off coal won't be easy. China promises to cap its emissions by 2030. It is likely to become 40 per cent more energy efficient than 2015.

The country has also committed to boosting its forest cover, and it is experimenting with a carbon trading plan: Lawmakers recently approved a pilot program in seven provinces and cities.

European Union

The EU and its 27 member states have pledged to reduce emissions by 20 percent below 1990 levels by 2020. The EU has said it would bump this commitment up to 30 percent if other developed countries sign up for similar commitments.

Under the Kyoto Protocol, the then-15 EU member states signed on to reduce emissions by 8 percent below 1990 levels by 2012. To meet that goal, in 2005 the EU launched the biggest carbon trading market in the world. Today, all 27 member states are required to participate, plus Iceland, Liechtenstein and Norway. Major factories and power plants in the EU are granted permits for how much carbon they can emit. Companies that emit less carbon than their allotted amount can sell their extra carbon credits to firms that exceed their emissions limit.

Starting in January, all airlines with flights that take off or land in Europe will be required to buy carbon permits to offset emissions from their flights. That requirement has sparked objections and legal challenges from several nations that argue it violates international law.

India

India, like China, also **won't commit to reducing its carbon emissions** — saying that would hurt efforts to bring millions of its citizens out of poverty. But agreed to increase its energy efficiency by 20 percent by 2015.

India is the world's No. 3 emitter of greenhouse gases, but because it's a developing nation, it isn't required to cut emissions under the Kyoto Protocol. That said, India is an active participant in the Clean Development Mechanism — a carbon offset plan set up under the Kyoto Protocol. India has hundreds of CDM projects; almost half of them focus on wind power and biomass.

India has set an ambitious goal of getting100 gigawatts of solar power online by 2022. A gigawatt of electricity is enough to power a small city. In 2010, the country started levying a carbon tax on coal to help subsidize renewable energy projects.

Indonesia

Indonesia has pledged to cut emissions by **26 percent by 2020 from today's levels**.

Indonesia is home to vast swaths of tropical forests, which suck up atmospheric carbon. But those forests are being logged at an alarming rate — and that's releasing huge amounts of carbon into the atmosphere. Under a deal with Norway that went into effect in May 2011, Indonesia agreed to implement a two-year moratorium on new concessions for clearing forests in exchange for $1 billion in support for its forest conservation efforts.

But many observers question Indonesia's commitment to prevent deforestation, given that the country's current economic boom has been largely fueled by extraction of its natural resources. Allegations that Forestry Ministry officials have lined their political war chests with funds raised by selling off logging rights haven't done much to bolster confidence.

Japan

Japan has pledged to reduce its emissions by **25 percent below 1990 levels by 2020**.

The world's No. 5 greenhouse gas producer, Japan committed to reducing its emissions by 6 percent below their 1990 levels under the Kyoto Protocol, and it was largely on track to meet that goal. In 2010, it launched a cap-and-trade plan aimed at forcing some 1,300 major businesses — including large office buildings, public buildings and schools — in the Tokyo metropolitan region to reduce their emissions.

However, the Fukushima nuclear disaster threw Japan a fastball. The nation relied on nuclear power for about a third of its electricity, but in the wake of the March 2011 accident, the vast majority of its reactors have gone offline. The lost output forced Japan to institute energy-reducing measures and, in the short term, to rely more heavily on fossil fuel-burning power utilities — which boosted its emissions in 2011. With the Japanese public now wary about nuclear energy, the nation's leaders are trying to find a new way forward.

Russia

Russia has pledged to reduce its emissions by at least **15 percent from 1990 levels** — a year when the Soviet Union was still in existence, and emissions from heavy industry, mostly related to the military, were sky high.

When Russia ratified the Kyoto Protocol in 2004, it pledged to hold its greenhouse gas emissions at or below 1990 levels. After the Soviet Union collapsed, Russia's emissions did, too. So the country hasn't had to do much to meet its Kyoto pledges.

Indeed, Russia has long been known as a country with little regard for environmental concerns, and it is still largely dependent on many heavy industries that are considered major polluters. Despite Russian ratification of the climate pact, for a long time the country's leaders continued to question the human role in climate change.

In 2009, the Russian government quietly reversed that position, adopting a new climate doctrine that seemed to accept human contribution to global warming. Same year, the country pledged to reduce its emissions by at least 15 percent from 1990. However, this pledge still doesn't require any action on Russia's part: By some estimates, the country's emissions remain more than 30 percent below 1990 highs. Though Russia has unveiled energy-efficiency goals, analysts call the country's climate policies "a black hole."

South Africa

South Africa expects its **emissions to peak between 2020 and 2025**, then remain flat for a decade before dropping off. By 2020, South Africa aims for emissions to top out at levels 34 percent lower than if the country were to take no actions.

South Africa is highly dependent on coal — about 90 percent of its electricity comes from burning the fossil fuel — and it's a major contributor to greenhouse gas emissions in Africa. The nation is slowly studying cleaner energy options and more energy-efficient alternatives. But to move forward with any emission reductions, South Africa says that there is a need of need funding and support from industrialised nations.

South Africa's renewable energy initiative aims to make clean power account for nearly 9 percent of the nation's energy mix by 2030. But that project is just getting off the ground: Construction on the first few dozen projects, mostly wind and solar power plants, commenced after mid-2012.

The country says it's committed to make nuclear power — which currently supplies about 5 percent of its electricity — a much bigger part of its energy mix in the future. But a shortage of funding may delay those plans.

United States

The U.S. pledged to **reduce emissions by 17 percent by 2020**, but that promise was contingent on Congress passing an aggressive cap-and-trade bill. Instead, the bill ended up in the trash, and **the US hasn't made it clear how it will meet its emission goals.**

The US has taken some actions at the federal level to curb emissions, including new nationwide fuel-efficiency standards for cars and light trucks. Individual states also have laws designed to lower their emissions in the coming decades. California has the most ambitious plan: starting in 2013, the state will cap greenhouse gas emissions from factories and power plants, and, eventually, emissions from vehicles.

But even with all those state and federal actions taken together, the World Resources Institute figures that the U.S. can't achieve a 17 percent reduction in emissions by 2020. New federal laws — for example, one that puts a tax on carbon emissions — would need to fill the gap, and prospects for that aren't good.

———

13

PARIS AGREEMENT, 2015, ON CLIMATE CHANGE

Close to 200 nations' delegates, collected at Paris, France, to formulate "an Agreement" to halt the ruinous and irrevocable consequences, resulting from 'Global Warming', the cause of Climate Change. 150 Heads spoke on the November 30, 2015, at the inauguration ceremony, of their resolve to do everything in their power to save this earth from ruin; though each of them, also, highlighted his/her country-group's concerns and preferences, each very diverse one.

The Conference of Parties 21 (COP 21), under the aegis of UNFCCC, continued from 30th November till 12th December, 2015 a day after the scheduled date of closure. With strenuous efforts of heads of delegations (mainly ministers of the respective government), and the deft diplomacy of the French delegation, a consensus was arrived at. Each country had much to lose and much to gain. According to an expert of climate change, the world collectively has gained a lot at Paris; saved itself from its probable doom, a new beginning on the path of progress is envisaged. Every country gave in, some of its benefit, to arrive at the compromise.

i) Preparation of Agenda for Paris Summit

Conference of Parties- in LIMA (Peru) MEETING on December 1, 2014. (Excerpts from Indian Express December 1, 2014, as reported by Amitabh Sinha).

The differences among Conference participating countries to evolve Kyoto II Protocol (Paris Treaty) to bind countries to commitments on Climate Change (control) measures from 2015 onwards. The forthcoming meeting is the 20th. COP meeting of the series. The Kyoto II is to be finalised by February 2015 at the Heads of Nations meeting in Paris.

What is Being Negotiated

i) Mitigation or Emission Reduction: Must ensure mandated and targeted reduction in emission of greenhouse-GHG- gases. This issue is termed Intended Nationally-Determined Contributions, INDCs.

ii) Adaptation: Evolve action plans to help countries in adapting to the effects of climate change.

iii) Transfer of Technology. A framework to ensure easier access of technology, from advanced countries, to help reduce emissions or adaptation.

iv) Finance: A Technology Finance Fund was created, at Cancun COP, to help poorer countries.

Note: Some countries have already declared their commitments to emission reduction. The biggest hurdle, presently, is the 'legally binding' mitigation levels by all countries, including those in Annexure II of the Kyoto Protocol I agreed to in 1997; which permitted these countries to Common But Differentiated Responsibility and Respective Capability, (CBDR/RC). Annexure II countries did not have 'legally binding' commitments.

Agenda at Lima. The countries will negotiate on the basic elements and hopefully come to conclusions to provide the framework for Kyoto II Protocol.

What is in it for India? India is now the 4th largest, behind China, in GHG emissions. The envisaged rapid industrialisation and development programmes, will entail still higher emissions. Under Kyoto Protocol 1, India was not required to accept 'legally binding' mandates, but the world community is insisting on all countries to accept 'mandatory' cuts in GHG emissions. India is suggesting three way approach for itself.

The mandated reduction should be linked to an Index of "intensity of industrialisation". Financial Aid from the Technology Finance Fund. Transfer of Technology from advanced countries freely and on preferential terms.

2°C Temperature Rise by 2050

All countries accepted to limit temperature rise to 2° Celsius (C) by 2050 to limit effects of Climate Change To achieve this objective, the three major carbon emitters, USA, China and India should limit their emissions to 2-3 tons per capita by 2050 (Virtually impossible). India's present day emissions is less than 2 tons per capita (will rise due to major industrialisation).

Objective

The 2°C rise of temperature by 2050 was reiterated but the Paris Agreement officially pegged the rise to 1.5°C. Possibly, to retain a margin in hand. Though, the INDCs submitted at Paris indicated a predicted temperature rise of between 3°C and 4°C.

Negotiations For Agreement

i) Mitigation or Emission Reduction:- Must ensure mandated and targeted reduction in emission of greenhouse-GHG- gases. This issue is termed Intended Nationally-Determined Contributions INDCs.

ii) Adaptation: Evolve Action Plans to help countries in adapting to the effects of climate change.

iii) Transfer of Technology: A framework to ensure easier access of technology, from advanced countries, to help reduce emissions or adaptation.

iv) Finance: A Technology Finance Fund was created, at Cancun COP, to help poorer countries in the agreements in Kyoto II Protocol.

ii) The Paris Agreement

The principal features of Paris Agreement are elucidated below.

1. The Agreement is termed 'ambitious'. All countries will aim to limit temperature rise to 1.5 °C. The earlier (agreed) limit of 2°C of temperature rise (since industrial revolution) is the highest rise that can be tolerated. It means that the nations will reach "peak GHG, emissions" between 2060-2080". Thus, any more increase in temperature will be halted in these two decades.

   The bar was raised to save island nations from extinction. The foreign minister of Marshall Islands, Mr. Tony de Brum stated, "I can go back home to my people and say we now have a pathway to survival". People of low lying areas were terrified of going under sea waters within this century.

2. The Agreement is termed 'Equitable'. Every nation has submitted its GHG emission reduction/halting plan voluntarily; the plan is termed its "Intended National Development Commitment-INDC". There was no dictation or compulsion.

   187 nations handed over their INDCs to UNFCCC at Paris. These plans, collectively, will result in a temperature rise of between 2.7°C and 3.6°C; i.e. above the stipulated 1.5°C. To meet this commitment, there will be a five yearly "Global Stock Take" to assess and evaluate the effective implementation of each country's efforts to achieve its mitigation and adaptation targets. After the five yearly 'stock take' review, each country will be required to submit a fresh INDC; the revised plan will be more ambitious than that committed earlier by each country. The first 'Stock Take' is scheduled in 2018, and will be planned in every five yearly intervals thereafter.

3. Each developing country can continue with its economic development without restriction, except that it aids efforts to contain temperature rise of 1.5°C. However, the development programmes must help sustained ecological preservation.

4. All nations, not only rich nations listed in Annexure I of Kyoto Protocol, will contribute finance, according to their capacity, to the Green Climate Fund (GCF); though richer nations will be principal aid givers. The amount of 100 billion USD per year of GCF, agreed at Copenhagen COP in 2009, is not specifically ordained in the Paris Agreement.

   A significant addition of capital aid is envisaged from BRICS nations; albeit expected, because these nations will grow economically in the coming fifty years.

   Note: India believes that more than 200 billion USD GCF will be required annually by 2030.

5. There is no binding obligation on the advanced nations to transfer green / advanced technology to developing nations on easy financial terms. Nor the transfer will be free to least developed nations. Rich nations argue that private corporations own IPR in those technologies and they cannot be forced to forego their rights for free.

6. Hard bargaining took time on forcing developing countries to give up fossil fuels, coal in particular, for generating electricity. This issue was resisted strongly by India and other developing nations. No definite restriction was placed in the Agreement.

Observations and Comments by Experts and Others

1. The nations see climate change a stark reality. They virtually see its horrendous impacts looming large within foreseeable future. Therefore, the 'nay' sayers at earlier COPs became 'yes' players at Paris.

2. The lower temperature limit, 1.5°C, is set to help island states. They foresee their obliteration because of rising sea levels due to global warming. But developing nations, however, apprehend curtailment of their economic development plans and rights.

3. The responsibility of the rich and developed, Annexure I nations of Kyoto Protocol, for accumulation of GHGs and their life styles being prima facie, the cause of global warming there from, has been almost written off, at least verily diluted. The important clauses have the word "should" instead of "shall", thus releasing the erstwhile Annexure I countries from their obligations to compensate for polluting the planet.

   The principal of 'Common but Differential Responsibility, CDR, enshrined in the Kyoto Protocol, signed in 1997, has been given a go by. It may be recalled that according to the Kyoto Protocol, more than 30 rich and developed nations (USA, EU, Japan, Canada, Russia, Australia in particular) designated Annexure I countries, were held responsible for accumulation of CO_2 to unacceptable heightened levels in the atmosphere, since inception of the industrial revolution. Therefore, they were required to compensate the Annexure II nations with financial aid and transfer of green technologies. This exclusive obligation, the rich nations were not prepared to shoulder any more. A compromise principal, stated earlier, INDC, earned the consensus of nations and saved the Paris Conference from failure.

4. There is no firm commitment, on advanced nations, to contribute to GCF 100 billion USD by 2020, and more subsequently. This fund is now a universal responsibility of all nations.

5. The result of Paris deliberations is in two parts, a) non-binding and b) binding. The later remains to be drafted,

   Steps Towards A Green World

agreed to, and ratified by governments of at least 60 nations whose GHG emissions are equal to or more than 50% of the world's GHG emissions.

6. Several procedures and mechanisms to measure net emissions remain to be framed. These will determine, during five year reviews, the flexibility available to developing countries.

7. In short, the Paris Agreement will become the template for agreements on Climate Change in the future. The Rio Convention of 1992, and the Kyoto Protocol of 1997, both under UNFCCC, are history now.

8. In sum total, in the words of the foreign minister of France, Lauren Fabius, "Our collective effort is worth more than the sum of our individual effort." The Paris Agreement is a "Magnificent triumph," said Ban ki Moon, Secretary General of the United Nations.

International Solar Alliance (ISA), Inaugurated at Paris

Two Prime Ministers, Narinder Modi of India and Francois Hollande of France, launched the International Solar Alliance at Paris. What is it and Why?

It is a brilliant concept. There are 300 days of sunshine over 121 nation-countries, located between the Tropic of Cancer and the Tropic of Capricorn. Such huge energy resource must be harnessed by generating Solar Power; specific schemes are explained in Chapter 10.

Government Of India pledged, at Paris, to provide land, worth Rs. 400 crores, in Gurgaon near Delhi. A secretariat will be provided for the initial five years, and a building complex will be erected at the National Institute of Solar Energy at Gurgaon (Haryana).

Objectives of ISA

i) Promote solar technologies and investment with the aim to help poor communities in member countries.

ii) Promote R&D for solar technology, reduce the cost of capital as well as cost per unit of use.

iii) Create expert groups for (a) developing common standards, (b) testing, monitoring and verification protocols.

iv) Build teams to support efforts of member countries in establishing solar power projects, for electricity generations and heating purposes.

v) Build e-portal to share experiences.

If the projects under ISA programmes get implemented, then billions of the poor, of 121 nations can get access to electricity to upgrade themselves; with better agricultural practices and setting up industries like their richer counterparts.

———————

14

BEYOND 2016, NEW TARGETS, AUTHOR'S SUGGESTIONS

The Paris Agreement is a compromise; to avoid the stigma of "failure" of a summit attended by 150 Heads of States and 196 Nation States in all. There are a few suggestions, which if adopted, voluntarily, may ensure that each and every nation ensures that it reaches its maximum emissions between 2060-2080. We need ensure that this world will be worth living after 2100 A.D., a good one for our grand children and their progenies.

1. Every nation ratify the Paris Agreement without delay. Each country must place, in recognised institutions, 'National Legal Framework' to monitor, test and control adherence to their INCDs submitted at Paris. Report reduction of GHG emissions regularly to UNFCCC. Adopt and adhere to the five yearly reviews by UNFCCC.

2. All nations must contribute generously to Green Climate Fund, GCF. Rich and Developed nations ought to be generous (and gracious) to contribute to GCF. Neither the total contributions be limited to $100 billion per year; increased to $200 billion annually at the earliest. Nor, such contributions be made part of 'that country's Foreign Aid Outlay'. The management of the GCF must be strengthened; aggressive marketing and collection of contributions be a priority. Additionally, handing out support money to needy countries without delay (rule subservience) should be ensured.

3. Transfer of "Green Technologies" from the Developed to Developing and underdeveloped countries is a must. Rich countries must resolve "the Patent and Royalty" issues speedily. Let the latter contentious issues not hold up transfers. The "Patent and Royalty" is a financial issue primarily which can be settled between UNFCCC and the particular nation-country. The charges can be defrayed from the GCF in due time.

4. The share of solar power and electricity generation from (all) renewable resources must increase to 50% and above of each country's total power generation, by 2050. Promote 'smart grids'. Help of GCF and technology transfer be ensured.

5. Researches to replace 'fossil fuels' in an emergency, as also is "sequestering" CO_2 collected in the atmosphere. Long lasting, high storage capacity batteries, for transport vehicles, aircraft, and storing solar and wind power is another area of urgent research. (Rich countries may consider, with common consent of the US and Soviet parties, to place a pause on 'nuclear and 'cyber' researches).

6. Given below is a short list of "agreements", towards environment control, already in existence. These must be encouraged and given boost financially.

(i) Asia Pacific Partnership on Clean Development and Climate

The Asia Pacific Partnership on Clean Development and Climate is an agreement between Australia, China, India, Japan, South Korea, and the United States. The partnership had its official launch in January 2006 at a ceremony in Sydney, Australia. Within the past year, the six nations have initiated nearly 100 projects aimed at clean energy capacity building and market formation. Building on these activities, long-term projects are scheduled to deploy clean energy and environment technologies and services. The pact will be ineffective without any enforcement measures.

(ii) Federal Government Must Encourage the Lead By State Governments in USA

As of January 18, 2007, 8 Northeastern US states (Maine, New Hampshire, Vermont, Connecticut, New York, New Jersey, Delaware, Massachusetts. Maryland joined June 30, 2007) got involved in the Regional Greenhouse Gas Initiative (RGGI), which was a state level emissions capping and trading program. It is believed that the state-level program will indirectly apply pressure on the federal government by demonstrating that reductions can be achieved without being a signatory of the Kyoto Protocol. Observer states and regions: Pennsylvania, District of Columbia, Eastern Canadian Provinces, Rhode Island. On August 31, 2006, the California Legislature reached an agreement with Governor Arnold Schwarzenegger to reduce the state's greenhouse-gas emissions, by 25% by the year 2020. This resulted in the Global Warming Solutions Act which effectively put California in line with the Kyoto limitations. As of March 11, 2007, 418 US cities in 50 states, representing more than 60 million Americans support Kyoto after Mayor Greg Nickels of Seattle started a nationwide effort to get cities to agree to the protocol.

Large participating cities: Annapolis; Austin; Boston; Chicago; Denver; Lansing, Michigan; Little Rock; Los Angeles; Madison, Wisconsin; Minneapolis; New Orleans; New York; Philadelphia; Portland; Providence; Salt Lake City; San Francisco; San Jose; Seattle; Tacoma; West Palm Beach.

Hopefully, the Federal Government of USA will take note of these initiatives, and legislate a Federal Law to curtail GHG emissions at faster rate than those in its INDC submitted at Paris Summit, 2015.

(iii) Washington Declaration' Presidents or Prime Ministers from Canada, France, Germany, Italy, Japan, Russia, United Kingdom, the United States, Brazil, China,

India, Mexico and South Africa agreed, in January 2007, envisaged a Global Cap-and-Trade System that would apply to both industrialised nations and developing countries. This approach needs emphasis .

On June 7, 2007, leaders at the 33rd G8 summit agreed that the G8 nations would 'aim to at least halve global CO_2 emissions by 2050'. Hopefully, this sincere resolve will be steadfastly implemented.

7. Green Climate Fund to undertake 'researches' on the suggested below, measures:

i) Biotechnology which produces photosynthetic bugs that pump/throw out lots of energy. It would require billions in research to find out.

ii) Research in to crops that can withstand higher temperatures and vagaries of incessant rains or long dry periods.

iii) Cooling the earth by spraying particles in the stratosphere or by using salt to make clouds whiter, which can reflect back the radiations from sun.

iv) Clouds are losing ice content (which reflects back sun's radiations). This phenomenon must be halted. Otherwise, snow converted into water vapour, will let through more heat to earth's surface and thus, increase global warming.

> Note: scientist must ensure that "geo-engineering" projects do not create newer climate change issues and challenges.

(iv) Finally, humans ought to learn to withstand life on a hotter Earth, with lesser air conditioning and heated homes. The poorer nations ought to be helped.

8. Author's Views And Suggestions

1. Irrespective of what "others" do or do not do, it is imperative that India follows its own programmes, (which are quite ambitious). India should, however, fight for its

'views' aggressively. The 'developed nations' are not playing their role constructively; failing to provide "funds" and "transfer of green technologies".

2. Municipalities, RWAs and Individuals sincerely go all out to follow suggestions in Chapters 8 to11 .

3. The scary concerns are (a) Water famine, (b) Unpreparedness to face unexpected disasters

(i) **Shortage of Drinking Water**: There will be insufficient water for the Nation by 2025 (a prediction). Marathvada, Telengana, Andhra Pradesh, Gujarat, Madhya Pradesh, and Haryana are, already, facing water famine. Delivering it by tankers is not only inadequate, is at best, a short term measure:

(a) Rejuvenation of 'water bodies- ponds and 'all lakes' be taken up. Bizarre it may seem, consider releasing water from dams, even if it means reduction in generation of electricity.

(b) Rain water harvesting projects, community scale and individually should be "the priority".

(c) Drip irrigation agriculture to be encouraged (Follow practices observed in Israel).

(d) Desalination' of sea waters along the coastline, possibly to feed populations up to 100 km deep in land .

(e) Revise our cropping pattern. Reduce cultivation of sugarcane and rice definitely and possibly wheat too. Undertake growing of pulses and rough cereals which Indians must learn take up instead of rice and wheat.

(f) Agriculture research be driven towards finding "heat resistant" crops; stand higher temperatures or consume less water.

Humans cannot survive without water. Lives must be saved at all costs, be it losing comforts of electricity and, riding in spank automobiles.

Outlandish though it may appear, let experts weigh drilling of water to depths even up to those that is done for spotting 'gas or petroleum'.

(ii) **Disaster Management**: India and many other developing countries too, lacks adequate means to meet unexpected disasters, especially, excessive rains and floods on one hand and on the other hand provide water and food in draught hit regions, which are expansive. In short, "Disaster Management" needs up-gradation extensively. (a) Increase procurement of equipments and their supply, (b) train more people, and (c) train all citizens in self help, e.g. first aid, rescue minors and women.

4. A Group of Developing Countries/ Emerging Economies must join hands to rein in Global Warming. It is highly desirable, in their own interest, that the 10 highest carbon polluters act as a group; Group of Emerging Markets; among them Argentina, Brazil, China, India, Indonesia, South Korea, South Africa must collaborate. These countries must cut down uses of fossil fuels at the earliest. Non action or disjointed attempts will mean, besides the injurious effects on inadequate food and water availability etc.etc., the financial loss could vary from 2.2% to 4.5% of respective GDP

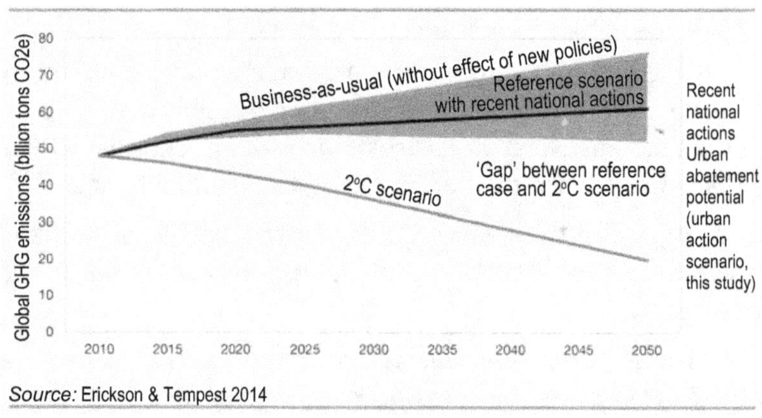

Source: Erickson & Tempest 2014

Fig. 14.1, There is potential for urban actions to reduce the emissions gap

Steps Towards A Green World

Role of Group of Emerging Market Economies (GMEs):

Strenuous efforts at rapid growth of economies by many developing countries, and globalisation of trade and finance will definitely lead to greater emissions of GHGs. Besides BRICS, there are more than 20 countries who are on the path of rapid growth. Among them are, Bangladesh, Czech Republic. Egypt, Indonesia, Iran, Turkey, Malaysia, Nigeria, Pakistan, Philippine , Poland, South Korea, Thailand, UAE, and many others in Africa and South America. The GMEs must act in concert to fulfill their commitments made at Paris in December 2015. In fact they must join hands with GDEs (Group of Developed Countries) to save the Earth. (Refer to chapter 15- Epilogue)

CONCLUSION. We have been warned. But, human beings always stand up to all types of challenges. They surely will overcome those that have been thrown up by Global Warming and the Climate Changes there from. The Steps Towards A Green World have been brought out in this book.

———————

15

EPILOGUE

1. India Ratifies Paris Agreement

India took the decisive step of 'Ratification of Paris Agreement on October 2, 2016. The Prime Minister said, "care and concern towards nature is integral to the Indian ethos. India is committed to doing everything possible to mitigate climate change".

The communication to the Chief of UNFCCC, says, "The government of India declares its understanding that, as per national laws, keeping in view its development agenda, particularly, the eradication of poverty, and provision of basic facilities of all its citizens, coupled with its low carbon path to progress, and on assumption of unemcumbered evailabilty of cleaner sources of energy and technologies and financial resources from around the world, and based on a fair and ambitious assessment of global commitment to combating climate change, it is ratifying the Paris Agreement".

Leaving aside the verbosity of the above notification, (no rich country has agreed to providing finance, or transfer of cleaner technologies) the government of India has set for itself the following:

By 2030, to achieve,

(i)   Reduce greenhouse gas emissions per unit of GDP by 30-35%.

(ii) At least 40% of electricity generation from non-fossil fuels.

(iii) Increase forest cover to create additional carbon cover of 2.5-3 bn. tons.

Some additional targets, but not part of commitments under Paris Agreement.

(iv) 100 GW of solar-installed capacity by 2022.

(v) 100 GW of wind-installed capacity by 2022.

(vi) 15 GW of biogas-installed capacity.

India's ratification is likely to hasten implementation of Paris Agreement. The agreement can come into force on the 30th day after the minimum of 55 countries, whose GHGs emissions aggregate 50% of the global emissions, ratify the agreement. To-date, the emissions of those who have ratified the agreement comes to 51.89%. Hopefully, Paris Agreement will commence by November 7, 2016, when COP takes place in Morocco. It is likely, either EU or Russia will ratify the agreement. Or any other combination of countries will make up the deficiency, appromiately 3% of global emmissions.

(Extracts from, Indian Express October 3, 2016)

2. Role of Developed Economies and Emerging Economies

Ward J. et al (2012) presented their findings (Self

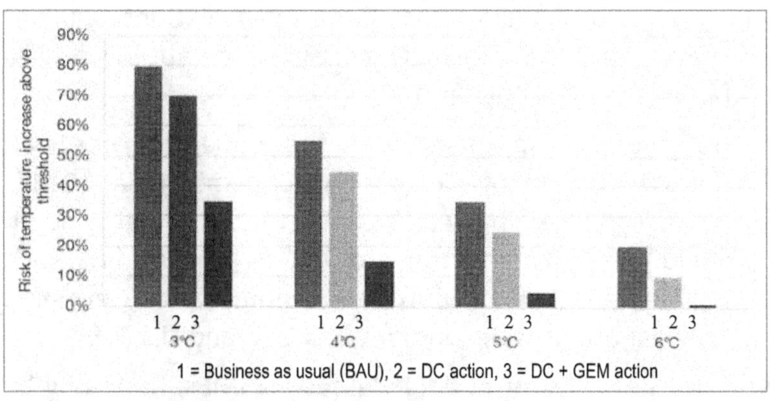

Fig. 15.1, GEM action is important to reduce the risk of catastrophic climate damage

-interested low-carbon growth in Brazil, China, and India Global Journal Of Emerging Market Economies, 4(3), 291-318). Some findings and suggestions are given below.

i) that the temperature rise, by 2100, will be in the range of 3.3°C (80% probability) to 6°C (20% probability) in case only Developed Countries convert promises to actions . The mean sea level rise will be 0.4m. to 0.5 m.

ii) BRICS Group will be the major defaulters, in view of their economic development programmes.

iii) Though India will be the major contributor of higher emissions, it by itself cannot halt the degradation, in spite of its very ambitious targets; some seemingly impossible to achieve without foreign financial and technology support.

iv) It is highly desirable, in their own interest, that the 10 highest carbon polluters act as a combine of Group of Emerging Markets; among them Argentina, Brazil, China, India, Indonesia, South Korea, South Africa must collaborate. These countries must cut down uses of fossil fuels at the earliest. Non action or disjointed attempts will mean, besides the injurious effects on inadequate food and water availability etc., the financial loss could vary from 2.2% to 4.5% of respective GDP.

v) The Author believes that Developed Nations will not change their Life Styles. Nor they will accept any legally binding restrictions on them.

However, hoping against hope, the cooperation between GDEs** (Group of Developed Economies) and GMEs (Group of Emerging Markets) will help in bringing down steadily the CO_2 emissions to 90% of present day levels (** Australia, Canada, European Union, France, Germany, Italy, Japan, Russia, Saudi Arabia, U.K. and U.S.A.).

vi) The Life styles of middle classes in GMEs, hopefully by

2050, will grow equivalent to that of middle classes in rich countries.

3. Solar Energy- World Bank Loan To India

The World Bank on Thursday (June 30,2016) announced a loan, of $ 1 billion in support of India's ambitious solar generation plans, its largest financing of solar projects for any country in the world. The projects now under preparation include solar rooftop technology, infrastructure for solar parks, bringing innovative solar and hybrid technologies to the market, and transmission lines for solar-rich States.

The commitment includes an agreement for a $ 625 million grid-connected rooftop solar programme for financing the installation of at least 40 megawatts of solar photovoltaic installations.

"We are doing all we can to support Prime Minister Modi's personal commitment to renewable energy, especially in scaling up solar energy," Dr. Kim said at a media briefing.

"India's plans to virtually triple the share of renewable energy by 2030 will both transform the country's energy supply and have far-reaching global implications in the fight against climate change," he said. According to him, India had become a leader in implementing the promises made in Paris for COP21 and the efforts against global warming.

The World Bank Group president also signed an agreement to be the financial partner for the International Solar Alliance that aims to collaborate on increasing solar energy use around the world, with the goal of mobilising $1 trillion in investments by 2030.

Note:- Dr. Kim, calls himself a "big fan" of the Prime Minister, Mr. Modi, "he always pushes us to move faster and faster – to keep pace with him".

(Source: News Papers)

4. Advancing U.S.-India Global Leadership on Climate and Clean Energy

Mr. Barack Obama, the US President and Mr. Narendra Modi, the Prime Minister of India, met in Washington on June7, 2016. Given below is the gist of the agreements concerning the climate Change. (Some parts are reproduced from their Joint Statement. Some part of this Statement is rephrased).

4.1 Clean Energies

i) The two Governments, through the U.S.-India Contact Group, addressed the nuclear liability issue, insisted by India. However, India's ratification of the Convention on Supplementary Compensation for Nuclear Damage, has laid a strong foundation for a long-term partnership between U.S. and Indian companies for building nuclear power plants in India. The leaders welcomed the start of preparatory work on site in India for six AP 1000 reactors to be built by Westinghouse of USA. India and the U.S. Export-Import Bank is to work together toward a competitive financing package for the project. Once completed, the project would be among the largest of its kind, fulfilling the promise of the U.S.-India civil nuclear agreement and demonstrating a shared commitment to meet India's growing energy needs while reducing reliance on fossil fuels. Both sides welcomed the announcement by the Nuclear Power Corporation of India Ltd, and Westinghouse that engineering and site design work will begin immediately and the two sides will work toward finalising the contractual arrangements by June 2017.

ii) The leaders, reiterated their commitment to pursue low greenhouse gas emission development strategies in the pre-2020 period and to develop long-term low greenhouse gas emission development strategies.

iii) The two countries resolved to work to adopt an HFC amendment in 2016 with increased financial support from donor countries to the Multilateral Fund to help developing countries with implementation, and an ambitious phase down schedule, under the Montreal Protocol pursuant to the Dubai Pathway.

iv) The leaders resolved to work together at the upcoming International Civil Aviation Organisation Assembly to reach a successful outcome to address greenhouse gas emissions from international aviation. Further, the two countries will pursue under the leadership of the G20 strong outcomes to promote improved heavy-duty vehicle standards and efficiency in accordance with their national priorities and capabilities.

v) The leaders welcomed the signing of 3 MOUs to Enhance Cooperation on Energy Security, Clean Energy and Climate Change on Cooperation in Gas Hydrates.

4.2 Clean Energy Finance

i) The United States supports the Government of India's ambitious national goals to install 175 GW of renewable power which includes 100 GW from solar power.

ii) To strengthen International Solar Alliance, the United States and India will jointly launch the third Initiative of the ISA which will focus on off-grid solar for energy access at the Founding Conference of ISA in September, 2016 in India.

iii) The United States also remains committed, with other developed countries, to the goal of jointly mobilising $100 billion per year by 2020 to address the needs of developing countries in the context of meaningful mitigation and adaptation action.

iv) The United States is committed to bring to bear its technical capacity, resources and private sector, and is jointly launching with India new efforts, to spur greater investment in India's renewable energy sector, including efforts that can serve as a model for other ISA Member Countries. In particular, the United States and India announced the creation of a $20 million U.S.-India Clean Energy Finance (USICEF) initiative, equally supported by the United States and India, which is expected to mobilise up to $400 million to provide clean and renewable electricity to up to 1 million households by 2020.

v) The two countries committed to establish the U.S.-India Clean Energy Hub as the coordinating mechanism to focus on increasing renewable energy investment in India; a $40 million U.S.- India Catalytic Solar Finance Program, equally supported by the United States and India, that, by providing needed liquidity to smaller-scale renewable energy investments, particularly in poorer, rural villages that are not connected to the grid, could mobilize up to $1 billion of projects; the expansion of handholding support to Indian utilities that are scaling up rooftop solar and continuation of successful cooperation with USAID on "Greening the Grid".

vi) The United States and India also remain committed to the goals of Mission Innovation, which they jointly launched during COP-21 in Paris to double their respective clean energy research and development (R&D) investment in five years. Toward this end, the leaders announced an upcoming $30 million public-private research effort in smart grid and grid storage.

Drinking Water Out Of Air (a technical innovation by Water-Gen Company of Israel)

The atmosphere is filled with water vapour; more vapour at higher temperatures (80°) that vapour is condensed into drinking water, which is very pure. The system is the best suited where water-pipe systems not available.

     Steps Towards A Green World

Governments which cannot layout costly pipe lines and need immediate solutions (armies deployed in mountainous and desert areas) can adopt such machines as described below:

Israel's company, Water Gen has developed machines of 3 sizes* which condense water vapour from the atmosphere. Water-Gen and six more companies are presenting at the United Nations General Assembly, their technological innovations this week (October 10-16, 2016 or thereabout).

Capacities:- 825 gallons, 118 gallons and 4 gallons per day, ; or 3122 liters, 446 liters, 15 liters,(1 gallon= 3.7 liters).

Machines, are highly suitable for hot and humid conditions, Latin America, Southeast Asia and Africa. Water-Gen is presently field-testing its products at Mumbai, Shanghai and Mexico City, besides many rural locations. This company's products are likely to be available here by next year. It has already sold its machines to several governments, including those of US, UK and France for use by their armies. Some Arab countries have, quietly, acquired Water-Gen machines.

Populations living in hot, humid, far-off places, unconnected with water-pipe supplies can very well obtain high quality drinking water. For once, governments neither need spend huge finances, nor wait long years to fulfill the basic human right to provide drinking water.

Amendment to Montreal Treaty- Phasing Out Of HFCs

Montreal Treaty was signed in 1989, phase out CFCs and HFCs has been modified a no. of times; yet it has proved the most successful international treaty. A hallmark amendment, to phase out HFCs (hydroflurocarbons), used for refrigeration was agreed to by 193 nations, on October 15, 2016 at Kigali, Rwanda, Africa. HFCs are several hundred or thousand times more powerful than CO_2 in inducing global warming. The elimination of these refrigerants by 2050 is estimated to prevent 0.5°C by the year 2100. The position, under discussion at Kigali, of USA, India and other developing countries was as under:

i)   USA insisted on the 'base year' 2011-2013 to commence depletion of HFCs.

ii) Brazil, China and many other developing countries proposed base year of 2020-2022.

iii) India, and some other developing nations, wanted the base year to be 2030. India had suggested its initial target of 10% reduction viz-a-viz 2024-2026, emissions. The reasoning is, that it requires more time to develop alternatives to HFCs without sacrificing growth of is industries engaged in preserving agricultural and horticultural products.

Finally, the negotiators of 193 nations agreed to the following, thus became an amendment to Montreal Treaty:

i)   India agreed to 2028 as 'freeze year'; the production and consumption of HFCs will not exceed its baseline period. A review of technology development will be undertaken in 2022. In case the alternative chemicals are not sufficiently developed the 'freeze year' will be revised to 2030. The target of reducing 10% HFCs by 2032 remains same, in both cases.

ii) The developed countries, led by USA, to start reducing HFCs by 2019 against a base line of 2011-13.

iii) Brazil, China and a host of other developing countries have the base line of 2020-22.

iv) HFCs to be eliminated completely by 2050.

The amendment to Montreal Treaty is likely to control the rising temperature; 0.5°C by the year 2100. This amendment thus, enables commencement of Paris Treaty within 30 days of this agreement; enhancement of commencement by 3 years or more.

16

DATA AND STATISTICS OF EMISSIONS

Note: The information in this chapter is culled from authentic sources. However, some of the data may not be the latest. All the same, the later data will show similar pattern/ relative to all countries.

A. Since the beginning of the Industrial Revolution, the increase in the concentrations of many of the greenhouse gases was only 50 ppm for 200 years, from 1770 to around 1973. Another increase of 50 ppm took only 33 years, i.e. to 2006; from pre industrialisation level of 280 ppm to 380 ppm in 2006.

B. Total Emissions of CO_2 in 2012, Giga Tons

1.

1.	China	10
2.	USA	5
3.	European Union	4
4.	India	2
5.	Russian Federation	2
	Total	23

2. USA 28%   China 16%   European Union 10%
   India 6%   Russia 6%   Japan 4%   All Others 30%

3. Highest Per Capita emission of CO_2 in 2012, Tons/per person

   1. UAE   23   2. Australia   19
   3. Saudi Arabia   18,   4. USA   17
   5. Canada   12   6. Russian Federation
                                  & Germany   10

India emits hardly, 1.7 t/p. There are a number of countries with much higher per capita emissions of CO_2

C. Carbon dioxide Emitting Countries and CO_2 Emission Per Capita- 2013

Source: Edgar data base, European Commission and Netherland Environment Assessment Agency includes CO_2 emissions "only" from fossil fuels and cement industry. Other than CO_2, No Green House Gas is included in the table.

Country	CO_2 emissions (1000 tons/year) in 2013	Emission per capita (tons per year) in 2013
World	35,669,000	5.0
China	10,540,000	7.6
United States	5,334,000	16.5
European Union	3,415,000	6.7
India	2,341,000	1.8
Russia	1,766,000	12.4
Japan	1,278,000	10.1
Germany	767,000	9.3
International Shipping	624,000	-
Iran	618,000	7.9
South Korea	610,000	12.3
Canada	565,000	15.9
Brazil	501,000	2.5
Saudi Arabia	494,000	16.8
International Aviation	492,000	-
Mexico	456,000	3.7
Indonesia	452,000	1.8
United Kingdom	415,000	6.5
Australia	409,000	17.3

South Africa	392,000	7.4
Turkey	353,000	4.7
Italy	337,000	5.5
France	323,000	5.0
Poland	298,000	7.8
Taiwan	277,000	11.8
Thailand	272,000	4.0
Ukraine	249,000	5.5
Spain	242,000	5.1
Kazakhstan	236,000	14.2
Malaysia	227,000	7.5
Egypt	225,000	2.7
United Arab Emirates	201,000	21.3
Argentina	199,000	4.8
Venezuela	195,000	6.3
Vietnam	190,000	2.1
Pakistan	158,000	0.9
Netherlands	158,000	9.4
Algeria	141,000	3.5
Iraq	139,000	4.0

D. Installed Power Generating Capacity in INDIA

Installed generating capacity, sector-wise and type-wise break up, as on August 31, 2016, is given below, in GW.

Sector/Type>	Fossil Fuels (Thermal)	Nuclear	Hydro	Renewable + Others	Total of a Sector
Central	58.9	5.78	11.65	0	76.33
State	71.8	Nil	28.2	1.96	101.96
Private	81.9	Nil	3.12	42.27	127.29
All India	212.6	5.78	42.97	44.23	305.58

The planned additional thermal power generation capacity excluding renewable power during the last two years of the 12th plan period (up to March 2017) is nearly 84,000 MW.

The Government of India has announced installation of 100 GW of electricity by renewable sources, includes Solar power (photovoltaic and thermal),d Wind power, Bio-Power, Co-generation from Bagasse, and Waste to Energy. It is visualised that renewable sources will contribute 40% of electric power by 2040. A welcome news is the competitive cost of electricity from Solar Energy.

The break-up of installed power of Renewable + Others was as below, in GW, on March 31,2015,

Small Hydro 4.055 + Wind 23.444 + Solar 3.744 + Bagasse cogeneration 3.008 + Biomass etc. 1.410 + Waste Power 0.115 = Total 35.776 GW.

E. Installed Power Generation of 11 Countries in GW (in 2008), Rank-wise

Country	Thermal Fossil Fuels	Nuclear	Hydro	Renewable Sources	Miscellaneous	Total
USA	3101	838	282	73	75	4369
China	2788	68	585	2.4	13.6	3457
Japan	711	258	83	10.4	19.6	1082
Russia	708	163	167	NA	NA	1040
India	685	15	114	16	NA-	830
Canada	162	94	383	8.5	3.5	651
Germany	388	148	72	29	NA	637
France	55	439	68	5.9	7	575
Brazil	59	14	370	20	NA	463
South Korea	288	14	151	20.6	NA	NA
U.K.	310	52	NA	NA	NA	389

F. Production of Electricity of 11 Countries in GWh. Rank-wise

China	5,810,500 GWh	Year 2015	Rank 1
USA	4,297,300	2014	2
E.U.	3,166,000	2014	3
India	1,208,400	2014	4
Russia	1,064,100	2014	5
Japan	1,061,200	2014	6
Canada	615,400	2014	7
Germany	614,000	2014	8
Brazil	582,600	2014	9
France	555,700	2014	10
South Korea	517,800	2014	11
Global	23,536,500	2014	NA

G. Solar Energy Facts (Amitabh Sinha Indian Express - 6-01-2015)

1. 15 GW total installed capacity in early 2014; end 2013-India plans to increase Solar Power to 100 GW by 2022.
2. 4 GT Carbon dioxide emissions avoided annually.
3. $ 96 BN new investment in PV Capacity in 2013.
4. 1.4 million jobs created globally by PV industry; 112,000 in India, 300000to 500000 in China.

H. Highest Temperatures Recorded

World	56.7°C Ferns Creek, California, USA recorded on 10-07-1913
Asia	53.5°C Mohenjodaro, Pakistan recorded on 26-05-2010
India	50.6°C Alwar, Rajasthan recorded on 10-05-1956
Delhi	47.8°C Recorded on 8-06-2014

Note: Excerpts from "THE HINDUSTAN" (published in Hindi at Delhi) dated 28-05-2015

I. Water Resources

1. Fresh Water is only 3%, the remaining 97% is Saline Water, resting in oceans.

2. Fresh water is used by humans for drinking, household chores, industries, agriculture and environment etc.

   Fresh Water distribution is 66.7% is in Glaciers and Ice-caps, 30.1% is Ground Water, 0.3% is Surface Water, and 0.9 % is in all Other Forms.

   Surface Water, 87% is in lakes, water bodies etc. 11% in swamps and only 2% in rivers.

3. Israel and India jointly organised a 4th India Water Conference at Delhi, from 4th. April, to 8th April 2016. The following information, annual per capita water in Cu.M availability, was released.

   i) Norway 8034   ii) *Russia   iii) *Australia   iv) *Brazil
   v)* USA   vi) South Africa   vii) Egypt 2037
   viii) China 419 ix) 209   x) Bangladesh 141   xi) Ethiopia 67
   Figures not known to Author

4. 4000, Giga Cu. M rain water annually is washed away.
   (As reported in Hindustan (Hindi) of 11.4.2016)

PART II

Basics of Climate Forming from Pre-historic Days

17

DEFINITIONS OF WEATHER, CLIMATE AND CLIMATE CHANGE

The terminology used in this book is defined and explained by meteorologists and scientists. This description will also aid understanding of the subject matter contained in Part I of this Book.

Weather

Weather is the day-to-day state of the environment and atmosphere. Ordinarily and mostly a common man thinks of weather in terms of sunshine, clouds, rain and wind-speed at a specific time and location. Most known parameters are temperature (max. and min.), relative humidity, and precipitation. On rainy days the amount of rain (measured according to standards universally accepted) is stated. Pollution watch is a recent addition to weather reporting.

Meteorological Departments, in each country the world over, collect and exchange data of numerous variables, e.g. Atmospheric Temperature (max. and min.), Humidity, Precipitation, Humidex, Dew point, Water vapour, Wind speed and direction, Wind chill, visibility distance Convection, Vorticity, Sea surface temperature, Heat index, Surface solar radiation, Visibility, Baroclinity, Cloud, Lightning, Storms and Hurricanes. Presence of fog (at important airports in particular), Pollution watch, presence of suspended matter particles is common these days. There may be too many more variants to be counted and enumerated here.

The data collected by weather balloons, weather satellites, automatic weather stations and buoys, is reported hourly to METAR or every 6 hours to SYNOP. These bodies and AMDAR (data collected through commercial airplanes) distribute, worldwide, their recordings. In recent times, satellite surveillance has made it easy to simultaneously collect data at many sites The World Meteorological Organisation has standarised instrumentation and methods of measuremens. Details are too many, but of little interest to readers.

In brief, the Weather is a dynamically condition of numerous (mostly the afore-stated) parameters. These assist to predict weather conditions over a particular area at a particular period.

Foreknowledge of 'Weather' is of utmost importance to agriculturists and horticulturists; crops and fruit growers. Any likely changes in onset of seasons, monsoons in particular, are predicted months ahead of their arrival. 24 hourly forecasts are regularly made. Forecasts up to 5 to 6 days ahead, from time to time, are broadcast on radio and television. The print media also provide essential details. Pilots of aero planes and ships are constantly advised of conditions around and ahead of them. Long-distance travelers in cars, coast guards and even office goers are warned of any impending unfavourable weather at short notice.

At times the changes in weather are sudden, chaotic and in strange ways; within a day or even in a few hours, sunshine, thunderous clouds and rain or snowfall descend and vanish. Jokes abound about forecasts of weather issued by Meteorological Departments, because they turn out to be unreliable. But, the number of variables is very large and their interplay is complex. Certainty in weather forecasting is still a distant goal!

Climate

Climate (from ancient Greek "Clime") is commonly defined as the weather averaged over a long period of time. Climate in a narrow sense is usually defined(by International Panel on Climate Change, IPCC) as the "average weather", or more rigorously, as the statistical description in terms of the mean and variability of relevant quantities over a period of time ranging from months to thousands or even millions of years. The classical period is 30 years, as defined by the World Meteorological Organisation (WMO). These quantities are most often surface variables such as temperature, precipitation, and wind. Climate in a wider sense is the state of weathers, in terms of a statistical averaged description.

Modern Climate Records

Modern climate records are based on measurements from weather instruments, such as thermometers, barometers, anemometers and a host of special ones. The instruments are calibrated for their known errors, and their immediate environment. Since improvements have taken place, the past records must be modified for the purpose

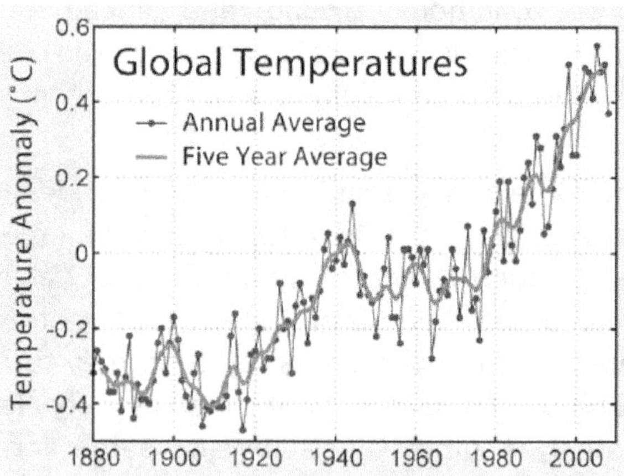

Fig. 17.1, The instrument recorded temperatures from 1850
[coloured version of the fig is on page 233]

of comparison. The practice to collect data on pollutants and greenhouse gases emissions has also been initiated. Computer modeling is employed to forecast climates in the distant future; even 25 to 100 years hence.

Climate's Determinants

Truly said, weather and climate include impacts of whatever is happening on the ground, below the ground as well as in the atmosphere above. Elaborating, any activity natural or man-made in the environment, e.g. atmosphere up to about 100 kilometers above, tectonic rocks lying underneath earth's surface, vegetation and plants, fauna and flora, animal species including humans, our foods and living styles and industrialisation, collectively impact and interplay among them influence weather and climate on the planet. Even the planet's spinning around itself, brings about changes in the environment, in turn influencing the natural process of weather shaping.

There are a number of static variables that determine climate, including: latitude, altitude, proportion of land to water, and proximity to oceans and mountains. Other climate determinants are more dynamic. For example (a) the 'thermo-haline circulation of the ocean' distributes heat energy between the Equatorial and Polar Regions, (b) many other ocean currents do the same between land and water on a more regional scale, (c) degree of vegetation coverage affects solar heat absorption, water retention, and rainfall on a regional level, (d) alterations in the quantity of atmospheric greenhouse gases and (e) determine the amount of solar energy retained by the planet.

The variables, which determine climate are numerous and the interactions complex. However, the broad outlines have been understood and agreed to by Meteorologists, in so far as the determinants of historical climate changes are concerned.

Seasons

The four popularly known seasons globally are, (i) Spring, (ii) Summer, (iii) Autumn and (iv) Winter. In India, according to folklore, there are six seasons, e.g. Winter-mild (Sheet), Spring (Vasant), Summer (Garmi), Monsoon (Sawan), Autumn (Sharad, Patjhar) and Winter-severe (Shishir). Equatorial regions have only *Dry-Hot, Dry- Cold and Wet seasons.

World Climate Classification

Originally the Greek's 'Clime,' was a concept used to divide the world into climatic zones sharing similar climatic attributes. Modern climatologists seek a close correspondence between climatic regions and vegetation biomes. Kopen scheme and the Thornthwaite climate classification are better known .

The Köppen scheme is one of the most widely used climate classification systems. It was developed by Wladimir Köppen, a German climatologist, firstly around 1900 and modifications by himself, notably in 1918 and 1936. According to this scheme, native vegetation is the best expression of climate. The climate zone boundaries, keeping vegetation distribution in mind, are based on average annual and monthly temperatures and the seasonality of precipitation. Köppen climate classification scheme divides the climates into 5 main groups, A to E. (Tropical, Arid, Temperate, Continental, and Polar). There are numerous subtypes, designated by a 2 to 4 letter suffixes, a, b, f, c, etc. Weathers can be reasonably derived from this climate classification. The Trewartha climate classification scheme modifies the Köppen system; it redefines the vegetation-zoning closer to the vegetation territories in the United States. (For Comprehensive Classification is described in the Appendix to this chapter.

(Note:-Thornthwaite classification is not given)

　　　　　　　　Steps Towards A Green World

Universal Thermal Scale

An option exists to include information on both the warmest and coldest months for every climate by adding a third and fourth letter suffix, e.g. - i for severely Hot and e for excessively cold, a, b, etc. for temperatures in between. Examples are Afaa for Kuala Lumpur, Malaysia, BWhl for Aswan, Egypt, Crhk for Dallas, Texas, DOlk for London, EClc for Arkhangelsk, Russia, and FTkd for Barrow, Alaska. (*The Universal Temperature Scale is given in the Appendix*)

Climate Change

Climate changes refer to variations of climate, globally or in well defined regions. These reflect changes within the Earth's atmosphere, oceans and ice caps, on account of natural causes and also the impact of human activities. The external factors that can shape climate are often called climate forcing and include such processes as variations in solar radiation, the Earth's orbit circling, and Greenhouse Gas concentrations; the most affecting cause of climate change. The climate changes or average state of the atmosphere can occur over time scales ranging from decades to millions of years. Evidence for climatic changes, in the past, is taken from a variety of sources that can be used to reconstruct olden climates, on the bases of modeling techniques . Climatic changes are inferred from changes in temperatures, vegetation, dendrochronology (the dating technique based on the investigation of annual growth rings in tree trunks) ice cores, sea level change, and glacial retreat.

Naomi Oreskes, a science historian, noted that "there's a huge disconnection between what professional scientists have studied and learned in the last 30 years, and what is out there in the popular culture". **In recent usage, especially in the context of environmental policy, the term "climate change" often refers generally to changes** in modern climate, including the rise in average surface temperature known as global warming. In some cases, the term, climate change, is also used with a presumption that human activities causation, as declared by the IPCC and United Nations Framework Convention on Climate Change (UNFCCC).

The UNFCCC also takes in to account non-human caused variations Climate change factors, which allude to Five natural sources, (glaciers, sun, soil, air and water which existed from the inception of this planet). The impact of these natural sources on climate shaping, are described in chapter 7 (Part II). Besides these, the effects of orbital revolving of the planet and volcanic eruptions are of significance in this context. In chapter 3, the Earth's Atmosphere is described, because most of the phenomena relating to Weather, Climate and Global Warming occur in the earth's atmosphere.

Climate Change in Popular Culture

The issue of climate change has entered popular culture since the late 20th century. Science historian Naomi Oreskes has noted that "there's a huge disconnect between what professional scientists have studied and learned in the last 35 years, and what is out there in the popular culture". An academic study contrasts the relatively rapid acceptance of ozone depletion as reflected in popular culture with the much slower acceptance of the scientific consensus on Global Warming. **Political advisor Frank Luntz recommended the Bush Administration adopt the term "Climate Change" in preference to Global Warming**, in fact there was a move to discredit the idea of global warming science.

All in all, the world has accepted the findings and conclusions, by IPCC and UNFCCC, that Global Warming is the primary cause of Climate Change (*Explained in Part I*).

Classification Of Climates

Glossary of Weather and Climate

Fluvio-glacial is the water created by the melting of glaciers. It literally means "Water Glacier." It is also commonly called 'melt water'.

Liquid water path [g/m^2] is a measure of the total amount of liquid water present in an air column. It is a retrieved quantity from the microwave radiometer.

Moisture ratio is a ratio that compares the mass or volume of air to the mass or volume of moisture contained in that air. In construction, it is an important consideration when designing a building for a certain climate. Nall (as cited in "References") called it one of "the most important climate variables for human comfort and building energy efficiency".

Hydrometeorology is a branch of meteorology and hydrology that studies the transfer of water and energy between the land surface and the lower atmosphere.

Cooperative Institute for Climate Applications and Research

The Cooperative Institute for Climate Applications and Research (CICAR) formalises a major collaborative relationship between the National Oceanic and Atmospheric Administration (NOAA) Office of Oceanic and Atmospheric Research (OAR) and Columbia University.

The CICAR research themes are:

- Modeling, understanding, prediction, and assessment of climate variability and change.
- Development, collection, analysis, and archiving of instrumental and Paleoclimate data.
- Development of the application of climate variability

and change prediction and assessment to provide information for decision makers and assess risk to water resources, agriculture, health, and policy.

Climate Monitoring and Diagnostics Laboratory

The Climate Monitoring and Diagnostics Laboratory (CMDL) was a climate laboratory in the National Oceanic and Atmospheric Administration (NOAA)/ Office of Oceanic and Atmospheric Research (OAR). In October 2005, it was merged with five other NOAA labs to form the Earth System Research Laboratory.

CMDL's mission was to observe and understand, through accurate, long-term records of atmospheric gases, aerosol particles, and solar radiation, the Earth's atmospheric system controlling climate forcing, ozone depletion and baseline air quality, in order to develop products that will advance global and regional environmental information and services.

Storms are classified by various names, such as • Thunderstorm • Tornado •Tropical Cyclone (Hurricane) • Winter storm • Blizzard •Low clouds, less than 1800 meters above sea level are called 'stratus' and 'Strato-cumulus' type .The clouds upto 6000 meters are named 'altostratus', 'altocumulus', and 'nimbostratus'. Those at highest levels have names like 'cirrus', 'cirrostratus' and 'cirrocumulus.'"

"These names are all Greek and Latin to us. Which one caused the cursed hurricanes, Katrina and Rita," asked Manu apologetically.

"They must have been 'cumulonimbus', which rise to towering altitudes and generally have rounded bumps on their undersides."

Precipitation can cause fog, drizzle, rain, freezing rain, sleet, hail, ice and snow.

Comprehensive Classification of Climates

GROUP	SUB GROUP and its PRINCIPAL CHARACTERISTICS T & P mean 12 month's average temperature and precipitation respectively
A Tropical/ Mega-thermal climates Tropical- primarily, sea level and low elevations. 12 months average temperatures 18 °C (64.4 °F) or higher	**Tropical Forest- Af**: 12 months average P is at least 60 mm (2.36"), within 5-10° latitude or extend up to 25° away from the equator on some eastern coasts. Sitiawan, Malaysia), Santos, Brazil, Singapore. **Tropical Monsoon Am**: Most common in Southern Asia and West Africa. The monsoon winds change direction according to the seasons. It has a driest month, soon after the "winter" solstice, P less than 60 mm. Konkary- Guinea, Chittagong-Bangladesh, Miami-Florida. **Savanna- Aw, As**: P less than 60 mm. The driest months, May, June Bangalore- India, Veracruz- Mexico, Townsville- Australia
B Dry (Arid and Semi-arid) Threshold P is decided by Agreed formulas	**Arid- BWh**: Threshold P is calculated by given procedure. Desert like, P, less than half the threshold, T above 18°C. Yuma, Arizona (BWh). T below 18°C P between half and one time threshold. Cobar, New South Wales, Australia, Murcia, and Spain. **Semi-arid**-BSh, BSk: As above, except one month temperature below 0°C, Turfan, Xinjiang, China, Medicine Hat, Alberta, Canada

C	Humid Subtropical- *Cfa, Cwa:* f means uniform rainfall, w means driest month in winter, P, <30 mm. s means driest month in summer. a means T, above 22°C, b means 4 months temperature above 10°C and T below 22°C and c means 3 or fewer months T, above 10°C. Luodian, Guizhou, China (*Cwa*). Houston, Texas, Brisbane, Queensland, Australia Yalta, Ukraine Porto Alegre, Brazil (*Cfa*)
Temperate/ Meso-thermal climates Summer T above 10 °C coldest month T from −3 °C to 18 °C.	**Oceanic**- *Cfb, Csa & Cwb:* Marseille, France Los Angeles, California, Perth, Australia (*Csa*). Limoges, France, Langebaanweg, South Africa Curitiba, Brazil (*Cfb*) which can be at higher altitudes also. Porto, Portugal San Francisco, California (*Csb*)
	Sub-polar Oceanic climates (*Cfc*): These climates occur pole-ward of the Maritime Temperate climates, and are confined either to narrow coastal strips on the western pole-ward margins of the continents, or, especially in the Northern Hemisphere, to islands off such coasts. Punta Arenas, Chile, Monte Dinero, Argentina, Iceland,
	Tórshavn, Faroe Islands, Reykjavík, Iceland, Tórshavn, Faroe Islands, Harstad, Norway
D Continental/ Microthermal T above 10 °C in warmest months, T below −3 °C (or 0 °C in USA) in coldest months	**Hot Summer Continental** (*Dfa, Dwa*): Dfa occurs in the high 30s and low 40s latitude. Lowell, Massachusetts Chicago, Illinois, *summer wetter than winter*, Santaquin, Utah, *summer drier than winter*
	Dwa extends further south in eastern Asia. Due to high pressure system of Siberia, winters are dry, and summers are wet because of monsoon. Seoul, South Korea

Warm Summer Continental (*Dfa, Dwa*): mostly in the 50s North latitude, although it might occur as far north as 70° latitude. In central and eastern Europe and Russia, it extends between high 50s and low end 60 degrees latitude. Moncton, New Brunswick, Canada, Minsk, Belarus, Revelstoke, British Columbia, Canada (*Dfb*) P can be uniform, or summer drier than winter or Fargo, North Dakota, winter drier than summer. Dsa exists only at higher elevations adjacent to areas with Mediterranean climates, such as Cambridge, Idaho and Saqqez in Iranian Kurdistan. Dsb generally exists at higher elevations, like Dsa, but at higher latitudes, chiefly in North America, (not in Eurasia) Washington State is one such location

Subarctic or Boreal (taiga) (*Dfc, Dwc, Dfd*): Quebec, Canada (Dfc) Anchorage, Alaska (*Dfc — summer wetter than winter*) Mount Robson, Irkutsk, Russia (*Dwc*). Kirkenes, Finnmark, Norway (*Dfc - summer wetter than winter*) Mount Robson, British Columbia, Canada (*Dfc — summer drier than winter*)

High-altitude Mediterranean (*Dsa, Dsb, Dsc*). Confined exclusively to highland locations near areas that have Mediterranean climates, and is the rarest of the three as a still higher altitude is needed to produce this climate. Zubacki kabao, Montenegro (*Dfsc* perhumide Mediterranean snow climate) another Example: Galena Summit, Idaho

E Polar T below 10 °C all 12 months	**Polar, Tundra climate (*ET*)** Warmest month has an average temperature between 0 °C and 10 °C. Canada, Russia Antarctica, ET is also at high elevations outside the polar regions, above the timber line; Mount Washington, New Hampshire
	Ice Cap climate (*EF*): T below 0 °C., Scot Base, Antarctica, Eismitte, inner Greenland or North Ice land **Alpine (*ETH*)**

Another classification system is the Thornthwaite climate classification, to align with US vegetation zonal patterns.

Appendix 17.2

Examples of climate change

Climate change has continued throughout the entire history of the Earth. The field of paleoclimatology has provided information of climate change in the ancient past, supplementing modern observations of climate.

1. Climate of the deep past,

2. Faint young sun paradox,

3. Snowball earth,

4. Oxygen Catastrophe

5. Climate of the last 500 million years,

6. Phanerozoic overview,

7. Paleocene–Eocene Thermal Maximum,

8. Cretaceous Thermal Maximum

9. Permo–Carboniferous Glaciation,

10. Ice ages,

11. Climate of recent glaciations

12. Dansgaard–Oeschger event,

13. Younger Dryas,

14. Ice age temperatures,

15. Recent climate,

16 Holocene Climatic Optimum,

17. Medieval Warm Period,

18. Little Ice Age,

19. Year Without a Summer,

20. Temperature record of the past 1000 years,

21. Global Warming,

22. Hardiness Zone Migration.

———————

18

EARTH'S ATMOSPHERE

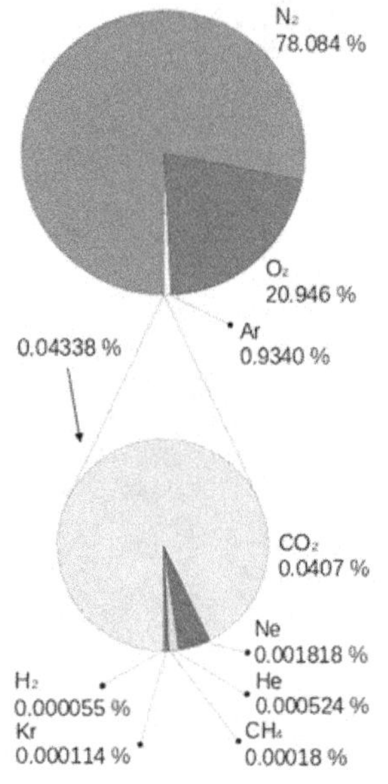

Nitrogen	78.084%
Oxygen	20.946%
Argon	0.934%
Carbon dioxide	0.038%
Water vapor	1%
Other	0.002%

Composition of air in December 1987(Others are shown in lower pie)

Earth's atmosphere
Troposphere	Stratosphere	Mesosphere	Thermosphere	Exosphere	Tropopause	Stratopause	Mesopause	Thermopause / Exobase	Ozone layer	Turbopause	Ionosphere

Fig.18.1, Layers of atmosphere - not to scale (NOAA)

Aerology studies the Ozone layer, atmospheric currents, climate, and environmental effects of the whole atmosphere. Methods include data collection by satellites, rocket-sondes, radio-sondes, weather balloons, and laser instruments. Analyses of the data help building models of predicting climates decades ahead.

Earth's Atmosphere

The atmosphere consists of layers of gases surrounding the planet Earth and held by the forces of gravity. The mixture of gases, though commonly known as air, contains other gases and particles of matter. The atmosphere protects life on Earth by absorbing ultraviolet solar radiation and maintaining minimum and maximum temperature extremes within limits over 24 hour cycles.

Air slowly becomes thinner and fades into space; breathing is difficult at higher altitudes. No definite boundary can be identified between the atmosphere and outer space. However, the Karman line, at 100 km (60 miles), is frequently accepted by scientists the higher end of the atmosphere. Three quarters of the atmosphere's mass is within 11 km (7 miles) of the Earth's surface.

In the United States, people who travel above an altitude of 80.5 km (50 miles) are designated astronauts. Space craft experience atmospheric effects at an altitude of 120 km (75 miles) during re-entry.

Atmosphere's Layers

Scientists have divided atmosphere in 6 layers; Troposphere, Stratosphere, Mesosphere, Thermosphere, Ionosphere and Exosphere which are described below.

1. **Troposphere**. The troposphere is the lowest layer of the atmosphere, extends between 0 to 7 km (0-4 miles)* at the poles and 17 km (0-10 miles) at the equator, with some variation due to weather factors. Solar heating of the earth's surface warms air masses. They rise to release latent heat as sensible heat that further uplifts

the air mass. Vertical columns keep on turning and mixing until all the water vapour is removed. Generally, temperature is lesser at heights due to expansive cooling. Temperatures in Troposphere vary considerably according to latitude and altitude of places.

Greenhouse Gases are mostly trapped in Troposphere.

The figures denote approximate distances above the Earth's surface, and are rounded off.

2. **Stratosphere**: The next atmospheric level lies above 7 -17 km level to about 50 km (30 miles). The concentration of Ozone is high from approximately 10 to 50 km (16- 30 miles). The Ozone layer's thickness varies seasonally and geographically. Ozone's concentration is lesser in the lower regions than the designated distances. However, ozone still forms only a low component, compared to other gases of the atmosphere. (Ref. chapter 5 for detailed discussion).

3. **Mesosphere**: From about 50 km (30 miles) to the range of 80 to 85 km (48-51miles), temperature decreases with height.

4. **Thermosphere**: From 80 – 85 km to 640+ km (400+ miles), temperature increases with height.

5. **Ionosphere**: Within Thermosphere, there are also ionised layers, electrically charged by solar radiation. Mostly, the phenomena of atmospheric electricity (lightning and thunderous clouds) happen within it. Radio waves' propagation is influenced by the ionosphere which also causes auroras; appearance of streamers of coloured lights in the sky near the earth's magnetic poles.

6. **Exosphere**: From 500 - 1000 km (300 - 600 miles) up to 10,000 km (6,000 miles), free-moving particles that may migrate into and out of the magnetosphere or the solar wind.

The boundaries between these regions are named the tropopause, stratopause, mesopause, thermopause and exobase.

The average temperature of the atmosphere at the surface of Earth is 15 °C (59 °F), highest being in 50°C in Africa's deserts and the lowest being minus 30°C in Siberia. The temperature of the Earth's atmosphere varies with altitude; the mathematical relationship between temperature and altitude varies among six different atmospheric layers.

Pressure and Thickness

Atmospheric pressure is the total weight of the air above the point of measuring pressure. It varies with location and time of measuring. The average atmospheric pressure, at sea level, is about 101.3 kilopascals (about 14.7 psi). Satellites experience atmospheric drag even in the exosphere, indicating presence of the atmosphere.

Well established mathematical equations of pressure by altitude are available to estimate atmospheric thickness. However, the following published data are given for reference.

50% of the atmosphere by mass is below an altitude of 5.6 km. 90% of the atmosphere by mass is below an altitude of 16 km. The common cruising altitude of commercial airliners is about 10 km. 99.99997% of the atmosphere by mass is below 100 km (almost all of it). The highest X-15 plane flight in 1963 reached an altitude of 108 km. In the rarefied region above this there are auroras and other atmospheric effects.

The mean molar mass of air is 28.97 g/mol. Note that the composition figures above are by volume-fraction (V%), which for ideal gases is equal to mole-fraction (that is, fraction of total molecules). By contrast, figures for mass-fraction abundances of gases, particularly for gases with significantly different molecular (molar) mass from that of air, will differ accordingly in their molar (mole) fraction abundance figures. For example, helium is 5.2 ppm by volume-fraction and mole-fraction, but only about 4/29 X 5.2 ppm = 0.72 ppm by mass-fraction.

Heterosphere

Below an altitude of about 100 km the Earth's atmosphere has a more-or-less uniform composition (apart from water vapour). However, in higher regions, the composition of the atmosphere varies with altitude. This is because, in the absence of mixing, the density of a gas falls off exponentially with increasing altitude. Thus higher mass constituents, such as oxygen and nitrogen, fall off more quickly than lighter constituents such as helium, molecular hydrogen, and atomic hydrogen. Hence, there is a layer, called the Heterosphere. As the altitude increases, the atmosphere is dominated successively by lighter gases. The precise altitude of the heterosphere and the layers it contains vary significantly with temperature.

Density and Mass

The density of air at sea level is about 1.2 kg/m^3 (1.2 g/L). Changing weather brings about variations of the barometric pressure at any one altitude. Solar radiations cause noticeable pressure variations in outer atmosphere, but hardly affect cities at lower altitudes.

Temperature and mass density vs. altitude, from the NRLMSISE-00 standard atmosphere model.

The average mass of the atmosphere is about 5,000 trillion metric tons or 1/1,200,000 the mass of Earth. According to the National Center for Atmospheric Research, "The total mean mass of the atmosphere is 5.1480×1018 kg with an annual range due to water vapour of 1.2 or 1.5×1015 kg depending on whether surface pressure or water vapour

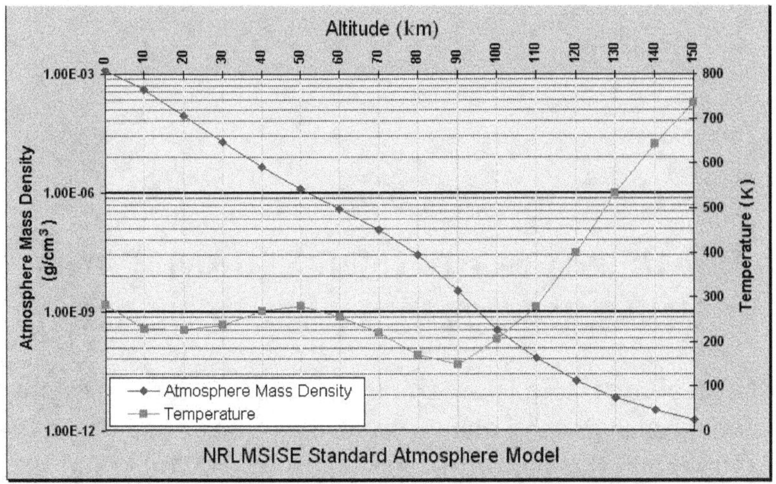

Fig. 18.2, Atmospheric mass density and temperatures against altitude [coloured version of the fig is on page 237]

data are used; somewhat smaller than the previous estimate. The mean mass of water vapour is estimated as $1.27×1016$ kg and the dry air mass as $5.1352 ±0.0003×1018$ kg."

The atmospheric density decreases as the altitude increases. This variation can be approximately modeled using the barometric formula. More sophisticated models are used by meteorologists and space agencies to predict weather and orbital decay of satellites.

19

BEAUFORT SCALE (FOR STORMS) AND UNIVERSAL TEMPERATURE SCALE

The Beaufort scale is an empirical measure for describing wind intensity based mainly on observed sea conditions. Its full name is the Beaufort wind force scale. The scale was created in 1805 by Irishman Sir Francis Beaufort, a British admiral and hydrographer. Initially, one man's "stiff breeze" might be another's "calm conditions". The scale from zero to 12 related qualitative wind conditions to guide the sails of the main ship of the Royal Navy, from "just sufficient to give steerage" to "that which no canvas [sails] could withstand." The scale was made a standard for ship's log entries on Royal Navy vessels in the late 1830s.

The scale was adapted to non-naval use from the 1850s, with scale numbers corresponding to cup anemometer rotations. In 1906, with the advent of steam power, the descriptions were changed to how the sea, not the sails, behaved and extended to land observations. Rotations to scale numbers were standardised only in 1923. George Simpson, Director of the UK Meteorological Office, was responsible for this and for the addition of the land-based descriptor. The measure was slightly altered some decades later to improve its utility for meteorologists.

The Beaufort scale was extended in 1946, when Forces 13 to 17 were added. However, Forces 13 to 17 were intended to apply only to special cases, such as tropical cyclones. Nowadays, the extended scale is only used in

Taiwan and mainland China, which are often affected by typhoons.

Wind speed on the 1946 Beaufort scale is defined by the empirical formula:

$$v = 0.836 \, B^{3/2} \text{ m/s}$$

where v is the equivalent wind speed at 10 metres above the surface and B is Beaufort scale number. For example, B = 9.5 is related to 24.5 m/s which is equal to the lower limit of "10 Beaufort".

Today, hurricanes are sometimes described as Beaufort scale 12 through 16, very roughly related to the standard Saffir-Simpson Hurricane Scale where Category 1 is equivalent to Beaufort 12. However, the Saffir-Simpson Scale does not match the extended Beaufort numbers above 13. Category 1 tornadoes on the Fujita and TORRO scales also begin roughly at the end of level 12 of the Beaufort scale but are indeed independent scales.

Note that wave heights in the scale are for conditions in the open ocean, not along shore.

The scale is used in the Shipping Forecasts broadcast on BBC Radio 4 in the United Kingdom.

This scale is also widely used in China. Taiwan uses the Beaufort scale extended in 1946 with Forces 13-17 to better represent the wind caused by typhoons. On the morning of May 15, 2006, mainland China suddenly introduced the extended scale to Force 17 without any prior notice. This extended scale was immediately put into use for Typhoon Chanchu. Hong Kong and Macau keep using Force 12 as the maximum and adopted a set of simpler descriptions for public. The descriptions used in Hong Kong are shown here.[clarify] Macau used similar descriptions except the term gentle is retained for Force 3.

In the United States, winds of Beaufort 6 or 7 result in the issuance of a small craft advisory, with force 8 or 9 winds bringing about a gale warning, 10 or 11 a storm warning (or "tropical storm warning" for 8 to 11 if related to a tropical cyclone), and anything to 12 a hurricane warning.

Beaufort Scale	Wind Speed Km/h	Description	Sea Conditions	Land conditions
0	0	Calm	Flat.	Calm. Smoke rises vertically.
1	1-6	Light Air	Ripples without crests.	Wind motion visible in smoke.
2	7-11	Light Breeze	Small wavelets. Crests of glassy appearance, not breaking	Wind felt on exposed skin. Leaves rustle.
3	12-19	Gentle Breeze	Large wavelets. Crests begin to break; scattered whitecaps	Leaves and smaller twigs in constant motion.
4	20-29	Moderate Breeze	Small waves.	Dust and loose paper raised. Small branches begin to move.
5	30-39	Fresh Breeze	Moderate (1.2 m) longer waves. Some foam and spray.	Smaller trees sway.

6	40-50	Strong Breeze	Large waves with foam crests and some spray.	Large branches in motion. Whistling heard in overhead wires. Umbrella use becomes difficult.
7	51-62	Near gale	Sea heaps up and foam begins to streak.	Whole trees in motion. Effort needed to walk against the wind.
8	63-75	Gale	High waves (2.75 m) with dense foam. Wave crests start to roll over. Considerable spray.	Light structure damage.
9	76-87	Strong Gale	Very high waves. The sea surface is white and there is considerable tumbling. Visibility is reduced.	Trees uprooted. Considerable structural damage.
10	88-102	Storm	Exceptionally high waves.	Widespread structural damage.

| 11 | 103-119 | Violent Storm | Exceptionally high waves. | Widespread structural damage. |
| 12 | 120 | Hurricane | Huge waves. Air filled with foam and spray. Sea completely white with driving spray. Visibility greatly reduced. | Considerable and widespread damage to structures. |

The Universal Temperature Scale

i — severely hot: Mean monthly temperature 35 °C (95 °F) or higher

h — very hot: 28 to 34.9 °C (82 to 95 °F) and above

a — hot: 23 to 27.9 °C (73 to 82 °F)

b — warm: 18 to 22.9 °C (64 to 73 °F)

l — mild: 10 to 17.9 °C (50 to 64 °F)

k — cool: 0.1 to 9.9 °C (32 to 50 °F)

o — cold: −9.9 to 0 °C (14 to 32 °F)

c — very cold: −24.9 to −10 °C (−13 to 14 °F)

d — severely cold: −39.9 to −25 °C (−39.9 to −13 °F)

e — excessively cold: −40 °C (−40 °F) or below.

Examples of the resulting designations include Afaa for Kuala Lumpur, Malaysia, BWhl for Aswan, Egypt, Crhk for Dallas, Texas, DOlk for London, EClc for Arkhangelsk, Russia, and FTkd for Barrow, Alaska.

Steps Towards A Green World

20
ROLE OF OZONE GAS AND OZONE LAYER IN CLIMATE CHANGE

Ozone Gas

Ozone is a gas, is an isotope of oxygen which has 3 oxygen atoms. It is, therefore, less stable than diatomic oxygen. Its name comes from the Greek word for smell (ozein). Ozone has the peculiar odour sensed during lightning storms, or that pervades in a wet fish market. To sense the ozone smell, its concentration must be between 0.0076 and 0.036 ppm. It turns into blue colour gas at atmospheric temperature and pressure, a dark blue liquid at minus112°C and into violet-black solid below minus 193°C. In good concentration it is present mostly in lower stratosphere. Otherwise, Ozone in low concentration permeates throughout the atmosphere and it is colourless.

Ozone's Uses

Because of the high oxidising properties of ozone, it is a high grade sanitiser and bacteria killer and has many consumer applications; such as,

i) Ozone machines, with or without ionisation, are currently used to sanitise (high ozone output) and deodorize non-inhabited rooms, ductwork, vehicles, boats, woodsheds, and buildings. Air purifiers, emitting low levels of ozone are, in use in the US hospital operating rooms to sterile the air.

ii) Ozonated water is used to launder clothes, sanitise

food, drinking water, and surfaces in the home. Studies at California Polytechnic University have proven that low levels of ozone dissolved in filtered tap-water can produce more than a four-log (99.99%) reduction in such food-borne microorganisms as salmonella, E. coli 0157 and others. Ironically, while ozone is considered an atmospheric pollutant (causes smog) by the US government, it can actually reduce pollutants like pesticides in fruits and vegetables.

iii) Ozone is used in spas or hot tubs with reduced levels of chlorine or bromine for keeping the water free of bacteria. To prevent cross-contamination among bathers, it must be used in conjunction with another sanitiser. Ozone is also widely used in treatment of water in aquaria and fish ponds. Its use can minimise bacterial growth, control parasites and removes or reduces "yellowing" of the water. As the Ozone rapidly decomposes, its controlled application has no effect on the fish.

iv) Currently, research is on to synthesise cyclic ozone molecules by hitting it with ultra-fast lasers. Creation and isolation of such cyclic molecules could allow more energy to be packed into rockets and hence may allow for farther space travel.

v) Ozone cannot be stored and transported like other industrial gases and must therefore, be produced on site. The dominating consideration is, cooler the water, the better the ozone synthesis. Therefore, very high quantities of cooling water flow are essential to dissipate heat of the process.

Ozone, an Air Pollutant

Ozone possesses disagreeable properties too. Low level ozone is regarded as a pollutant by the World Health Organisation. Though erroneously believed, it is not emitted directly by car engines or by industrial operations. Instead, it is formed by the reaction of sunlight on air containing hydrocarbons and nitrogen oxides. Thus ozone is formed directly at the source of the pollution or many kilometers down wind.

Ozone photolysis (separation of molecules by light) by UV light leads to production of the hydroxyl radical which removes hydrocarbons from the air. It is also the first step in the creation of smog; creates peroxyacyl nitrates in smog, which are powerful eye irritants. The atmospheric lifetime of Tropospheric ozone is about 22 days, the remnants are deposited at the ground.

There is a great deal of evidence to show that high concentrations (ppm) of ozone, created by high levels of pollution and daylight UV rays at the earth's surface, can harm lung function and irritate the respiratory system. Surveys show a connection between increased ozone caused by thunderstorms and admissions of asthma sufferers in hospitals. The World Health Organisation has issued guidelines for quality of 'air', based on levels of ozone which can cause measurable health hazards.

Ozone also forms specific cholesterol-derived metabolites that are thought to facilitate the build-up of fatty plaques-the cause of heart disease. Air mixed with Ozone interferes with photosynthesis and stunts overall growth of some plant species. There is also evidence of significant reduction in agricultural yields due to increased ground-level ozone.

Ozone as a Greenhouse Gas

Ozone adds to the greenhouse effect because it absorbs some of the infrared energy emitted by the earth. Quantifying the greenhouse gas potency of ozone is difficult, as it is not present in uniform concentrations across the globe. The most recent scientific review on the climate change (IPCC-Third Assessment Report) suggests that the radiative forcing of Tropospheric ozone is about 25% that of carbon dioxide.

Ozone Layer

Ozone in low concentration permeates the entire atmosphere. "Ozone Layer" is that part of the atmosphere where concentrations of ozone (O3) is "Relatively high" (a few parts per million). But it is still small as compared to the main components of the atmosphere. It is mainly located in the Stratosphere; from approximately 10 km to 50 km above Earth's surface. The thickness of the ozone layer—that is, the total amount of ozone in a column overhead— is in general smaller near the equator and larger towards the poles. It is thicker during the spring and thinner during the autumn. The reasons for these variations are complicated, involving atmospheric circulation patterns as well as solar intensity.

The ozone layer was discovered in 1913 by the French physicists Charles Fabry and Henri Buisson. Its properties

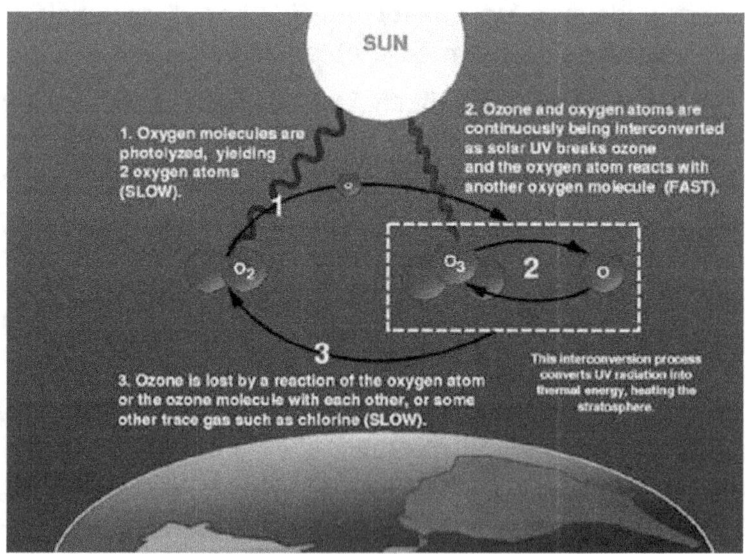

Fig. 20.1, Ozone-oxygen cycle in the ozone layer
[coloured version of the fig is on page 237]

Steps Towards A Green World

were explored in detail by the British meteorologist G.M.B. Dobson, who developed a simple spectrophotometer that could be used to measure Stratospheric ozone from the ground level. Between 1928 and 1958 Dobson established a worldwide network of ozone monitoring stations which continues to operate till today. The "Dobson unit", a convenient measure of the total amount of ozone in a column overhead, is named in his honour.

Ultraviolet Light and Ozone

The highest levels of ozone in the atmosphere are in the stratosphere. Here it filters out the shorter wavelengths (less than 320 nm) of ultraviolet light (270 to 400 nm) from the Sun that would be harmful to most forms of life in large doses; though vitamin D is also produced by the same wavelengths is good and essential for human health. Ozone in the stratosphere is mostly produced from ultraviolet rays reacting with oxygen. It is destroyed when it reacts with atomic oxygen.

In recent decades, the amount of ozone in the stratosphere has been declining. The emissions of CFCs

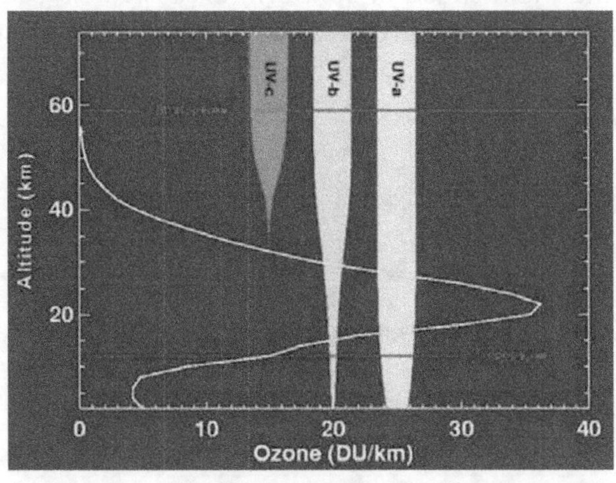

Fig. 20.2, Levels of ozone at various altitudes and blocking of ultraviolet radiation. [coloured version of the fig is on page 238]

(from air conditioners and refrigerators) and similar chlorinated and brominated organic molecules, have led to depletion of ozone because they increase the concentration of ozone-depleting catalysts above the natural background.

Levels of Ozone at Various Altitudes and Blocking of Ultraviolet Radiation

Ozone was present at ground level before the industrial revolution. Presently, its peak concentrations are far higher than the pre-industrial levels. Ozone's concentrations far away from sources of pollution are also substantially higher.

Although the concentration of ozone in the ozone layer is very small, it is vitally important to life because it absorbs biologically harmful ultraviolet (UV) radiation emitted from the Sun. UV radiation is divided into three categories, based on its wavelength; these are referred to as UV-A, UV-B, and UV-C. UV-C, which would be very harmful to humans, is entirely screened out by ozone at around 35 km altitude. However, it is interesting to note that ozone gas is a pollutant at lower levels and causes severe problems like oedema, hemorrage, etc. UV-B radiation can be harmful to the skin and is the main cause of sunburn; excessive exposure can also cause genetic damage, resulting in problems such as

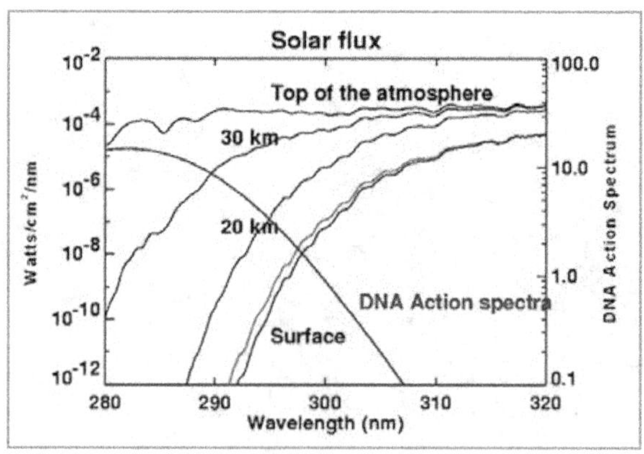

Fig. 20.3, DNA sensitivity to UV
[coloured version of the fig is on page 238]

skin cancer. The ozone layer is very effective at screening out UV-B; for radiation with a wavelength of 290 nm, the intensity at Earth's surface is 350 million times weaker than at the top of the atmosphere. Nevertheless, some UV-B reaches the surface. Most UV-A reaches the surface; this radiation is significantly less harmful, although it can potentially cause genetic damage.

Depletion of the ozone layer allows more of the UV radiation, and particularly the more harmful wavelengths, to reach the earth's surface, causing increased genetic damage to living organisms.

UV energy levels at several altitudes. Blue line shows DNA sensitivity. Red line shows surface energy level with 10% decrease in ozone.

To appreciate how important this ultraviolet radiation screening is, we can consider a characteristic of radiation damage called an action spectrum. An action spectrum gives us a measure of the relative effectiveness of radiation in generating a certain biological response over a range of wavelengths. This response might be erythema (sunburn), changes in plant growth, or changes in molecular DNA.

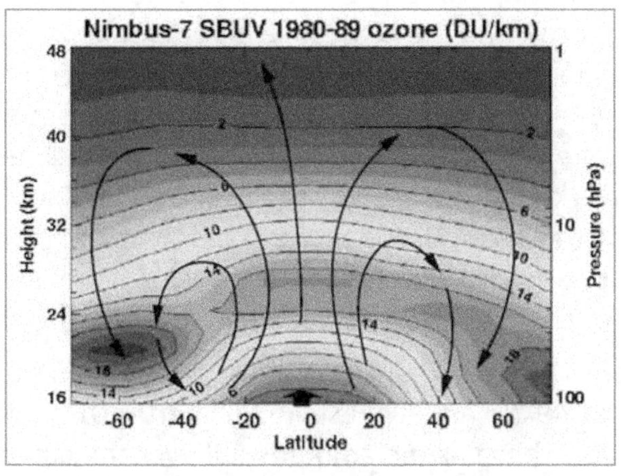

Fig. 20.4, Distribution of ozone in the stratosphere

There is much greater probability of DNA damage by UV radiation at various wavelengths. Fortunately, DNA is easily damaged only by wavelengths, shorter than 290 nm, UV which ozone strongly absorbs. At the longer wavelengths where ozone absorbs weakly, DNA damage is less likely. If there was a 10% decrease in ozone, the amount of DNA damaging UV increases by about 22%. Considering that DNA damage can lead to maladies like skin cancer, it is clear that this absorption of the sun's ultraviolet radiation by ozone is critical for our well being.

Total ozone concentration in June 2000 as measured by EP-TOMS in Nimbus satellite. The standard way to express total ozone levels (the volume of ozone in a vertical column) in the atmosphere is by using Dobson units (see below). Concentrations at a point are measured in parts per billion (ppb) or in $\mu g/m^3$.

Since, stratospheric ozone is produced by solar UV radiation, one might expect to find the highest ozone levels over the tropics and the lowest over Polar Regions. The same argument would lead one to expect the highest ozone levels in the summer and the lowest in the winter. The observed behaviour is very different: most of the ozone is found in the mid-to-high latitudes of the northern and southern hemispheres, and the highest levels are found in the spring, not summer, and the lowest in the autumn, not winter. During winter, the ozone layer actually increases in depth. This puzzle is explained by the prevailing stratospheric wind patterns, known as the Brewer-Dobson circulation. While most of the ozone is indeed created over the tropics, the stratospheric circulation then transports it pole-ward and downward to the lower stratosphere of the high latitudes.

Ozone amounts over the continental United States (25°N to 49°N) are highest in the northern spring (April and May). These ozone amounts fall over the course of the summer to their lowest amounts in October, and then rise again over the course of the winter. Again, wind transport of ozone is principally responsible for the seasonal evolution of these higher latitude ozone patterns.

Ozone Depletion

The ozone layer is depleted by free radical catalysts, including nitric oxide hydroxyl (OH), atomic chlorine (Cl), and atomic bromine (Br). While there are natural sources for all of these species, the concentrations of chlorine and bromine have increased markedly in recent years due to the release of large quantities of manmade organohalogen compounds, especially chlorofluorocarbons (CFCs) and bromofluorocarbons. These highly stable compounds are capable of surviving the rise to the stratosphere, where Cl and Br radicals are liberated by the action of ultraviolet light. Each radical is then free to initiate and catalyze a chain reaction capable of breaking down over 100,000 ozone molecules. Ozone levels, over the northern hemisphere, have been dropping by 4% per decade. Over approximately 5% of the Earth's surface, around the north and south poles, much larger (but seasonal) declines have been seen; these are called the ozone holes. (more details in Chapter on CFCs and Montreal Protocol).

21

ROLE OF CFCs & HFCs IN DEPLETION OF OZONE LAYER AND MONTREAL PROTOCOL

Fluorocarbons are chemical compounds that contain carbon-fluorine bonds, which tend to break up slowly in the environment. Therefore, many substances containing fluorocarbons are considered as persistent pollutants. CFCs also contain fluorine and chlorine atoms. They were used widely in industries as refrigerants, propellants, and cleaning solvents. However, CFCs generally have potent ozone-depleting potential. Their use has now been mostly prohibited by the Montreal Protocol (See below).

Uses and Applications of CFCs and HFCs and their Growing Concentration

Refrigerants: Some fluorocarbons (e.g. Freon) have been used as refrigerants. Because they deplete the ozone layer, many fluorocarbons have been banned as refrigerant after the Montreal Protocol.

Propellants: Many of these low boiling fluorocarbons were used as propellants before the Montreal Protocol.

Solvents: CFCs were used as industrial solvents, because they are non-flammable and very stable. Subsequent to the Montreal Protocol, HFCs were developed. Precision Cleaning (Degreasing), Defluxing of electronic assemblies, Particle removers, Driers after aqueous cleaning, as a Carrier fluid, and as a Dielectric coolant. HFCs can be tailored for specific applications; even beyond those mentioned.

Extracting Agent: HFCs, particularly, e.g. tetrafluoroethane, are used for extraction of extremely important natural products; such as Taxol for cancer treatment from yew needles, evening primrose oil food supplement, and Vanilla. Tetrafluroethane allows high quality and high yield extractions.

Lubricants: Fluorocarbons are non-reactive, low reactivity and very high temperature ranges. Solid fluoropolymers, as 'Greases' are used for demanding lubricating applications Teflon and other similar fluoropolymers are applied in layers to help reduce inter-layer-friction. Small, self-lubricated parts such as stopcocks for laboratory glassware may be entirely made of Teflon. Krytox, by DuPont is also used in certain firearm lubricants such as "Tetra Gun".

Water Repellant and Stain Repellant Products: Highly fluorinated organic compounds, e.g. Scotchgard containing fluorocarbons and perfluorooctane sulfonate (PFOS) have water-repellant and stain-repellant properties. But many of these uses have been phased out due to environmental concerns.

Pollution Effects

CFCs have been criticised for their harm to the ozone layer. It is estimated that a single CFC molecule has the ability to decompose approximately 1,00,000 ozone molecules.

Biological Flurocarbons

There are only a handful of natural fluorocarbons. They are found in microorganisms and plants, but not in animals. The most common natural fluorocarbon is fluoroacetic acid, a potent toxin found in a few species of plants. Others included ω-fluoro fatty acids, fluoroacetone, and 2-fluorocitrate which are all believed to be biosynthesised from fluoroacetic acid.

Montreal Protocol

The Montreal Protocol, an international treaty, was signed, at Montreal, Canada, on September 16, 1987 and came in to force on January 1, 1989. It is an international treaty designed to protect the ozone layer by phasing out the production of numerous substances, CFCs and HCFs, responsible for ozone depletion. It has undergone eight revisions, in 1990 (London), 1991 (Nairobi), 1992 (Copenhagen), 1993 (Bangkok), 1995 (Vienna), 1997 (Montreal), 1998 (Australia), 1999 (Beijing) and 2007 (Montreal).

As a result of the international agreement, the ozone hole in Antarctica is slowly recovering. Climate projections indicate that the ozone layer will return to 1980 levels between 2050 and 2070. Due to its widespread adoption and implementation, it has been hailed as an example of exceptional international co-operation. The two ozone (Vienna and Montreal) treaties have been ratified by 197 parties, which includes 196 states and the European Union, making them the first universally ratified treaties in United Nations history.

When comparing this very success story with futile attempts to establish an international policy on the Earth's climate change, Kofi Annan, former Secretary General of UN, while addressing a COP, of UNFCCC is quoted as saying, "Perhaps the single most successful international agreement to date has been the Montreal Protocol".

As a result of the international agreement, the ozone hole in Antarctica is slowly recovering. Climate projections indicate that the ozone layer will return to 1980 levels between 2050 and 2070.

Historical Background

In 1973, Chemists Frank Sherwood Rowland and Mario Molina, at the University of California, Irvine, discovered that CFC molecules were stable. They move upwards to the middle of the stratosphere. Finally, after an average of 50-100 years, the two common CFCs are broken down by ultraviolet radiation releasing a chlorine atom. Rowland and Molina proposed that these chlorine atoms might be expected to cause the breakdown of large amounts of ozone (O3) in the Stratosphere. Earlier on it had been established, that nitric oxide (NO) could catalyze the destruction of ozone. Though many other scientists had independently proposed that chlorine could catalyze ozone loss, but none had realized that CFCs were a potentially large source of chlorine. The environmental consequence of this discovery was that, depletion of the ozone layer by CFCs would lead to an in increase in UV-B radiation at the surface, resulting in an increase in skin cancer, damage to crops and to marine microscopic plants.

In 1976 the U.S. National Academy of Sciences (NAS), confirmed the scientific credibility of the ozone depletion hypothesis. NAS continued to publish assessments of CFC related researches for the next decade. Crutzen, Molina and Rowland were awarded the 1995 Nobel Prize for Chemistry for their work on CFCs.

On January 23, 1978, Sweden became the first nation to ban CFC-containing aerosol sprays that are thought to damage the ozone layer. A few other countries, including the United States, Canada, and Norway, followed suit later that year, but the European Community rejected an analogous proposal. Even in the U.S., chlorofluorocarbons continued to be used in refrigeration and industrial cleaning, until after the discovery of the Antarctic ozone hole in 1985. After negotiation of an international treaty (the Montreal Protocol), CFC production was sharply limited beginning in the year 1987 and phased out, by these countries, completely by the year 1996.

Montreal Protocol

The protocol states...Recognising that world-wide emissions of certain substances can significantly deplete and otherwise modify the ozone layer in a manner that is likely to result in adverse effects on human health and the environment, ... Determined to protect the ozone layer by taking precautionary measures to control equitably total global emissions of substances that deplete it, with the ultimate objective of their elimination on the basis of developments in scientific knowledge ... Acknowledging that special provision is required to meet the needs of developing countries...

Monetary Aid to Developing Countries

A Multilateral Fund was the first ever financial mechanism to be created under an international treaty. It embodies the principle agreed at the United Nations Conference on Environment and Development in 1992, that countries have a common but differentiated responsibility to protect and manage the global commons. (This very principle of a Green Climate Fund, is put forward by India to ease Global Warming).

The Fund provides financial aid to developing countries to phase out the use of ozone-depleting substances (ODS) and replace them by less harmful HCLs for refrigeration, foam extrusion, industrial cleaning, fire safety and fumigation.

The Fund is managed by an Executive Committee with an equal representation of seven industrialised and seven Article 5 countries which are elected annually by a Meeting of the Parties. An annual report on its operations is scruitinised and assessed by a Meeting of the Parties.

Up to 20 percent of the contributions of contributing parties can also be delivered through their bilateral agencies in the form of eligible projects and activities.

The Fund is replenished on a three-year basis by the donors. Pledges amount to US$ 2.1 billion over the period 1991 to 2005. Funds are used, for example, to finance the conversion of existing manufacturing processes, train personnel, pay royalties and patent rights on new technologies, and establish national Ozone Offices.

On August 2, 2003, scientists announced that the depletion of the ozone layer may be slowing down due to the international ban on CFCs. Three satellites and three ground stations confirmed that the upper atmosphere ozone depletion rate has slowed down significantly during the past decade. The study was organised by the American Geophysical Union. Some breakdown can be expected to continue due to CFCs used by nations which have not banned them, and due to gases which are already in the stratosphere. CFCs have very long atmospheric lifetimes, ranging from 50 to over 100 years, so the final recovery of the ozone layer is expected to require several lifetimes. Compounds containing C–H bonds are being designed to replace the function of CFC's (such as HCFC), since these compounds are more reactive and less likely to survive long enough in the atmosphere to reach the stratosphere where they could affect the ozone layer.

Schedule of Reduction and Elimination of CFCs, by Montreal Protocol.

It was declared that, Group I of Annex A of the protocol lists CFCl3 (CFC-11), CF2Cl2 (CFC-12), C2F3Cl3 (CFC-113), C2F4Cl2 (CFC-114), C2F5Cl (CFC-115) as the primary ODS, to be phased from 1991 to 2010. The less active HCFCs will be completely phased out by 2030.

Montreal Protocol Compliance

The overall level of compliance has been high. The atmospheric concentrations of the most important chlorofluorocarbons and related chlorinated hydrocarbons have either leveled off or decreased. In a 2001 report, NASA found the ozone hole over Antarctica had remained the same size for the previous three years. However in 2003, the ozone hole grew to its second largest size.

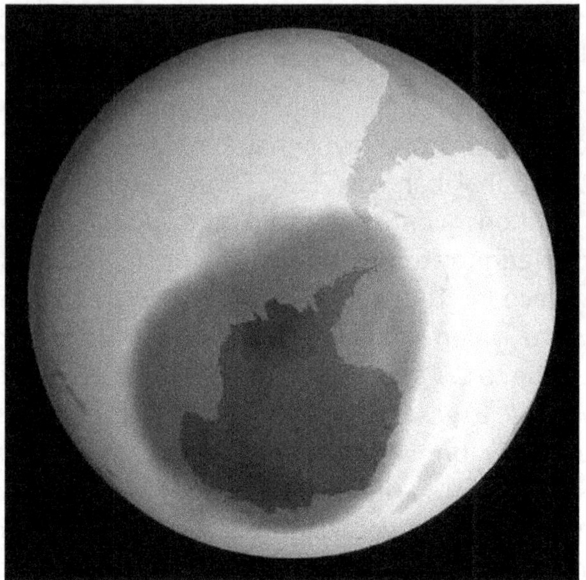

Fig. 21.1, Antarctic 'Ozone hole'- September 2000

The largest Antarctic ozone hole recorded as of September 2000. The general trends of Ozone Depleting Substances are shown in the figure above.

Steps Towards A Green World

22

EL NINO PHENOMENA

El Niño-Southern Oscillation (ENSO)

El Niño-Southern Oscillation (ENSO) ENSO. Ocean waters also warm up, because of sun's radiations. But differently in different parts, which set of interacting currents of water and air circulations. These circulations cause atmosphere climate fluctuations. ENSO is the most prominent known source of inter-annual variability in weather and climate around the world . ENSO primarily influences atmospherics in the Pacific, and Indian Oceans. The Southern Oscillation (SO) reflects the monthly or seasonal fluctuations in the air pressure difference between Tahiti and Darwin. In India, the most recent experience, deficient rainfall and draught in parts, was in 2015; such influence happens in intervals varying between 3 to 8 years; the earlier one occurred in 2006-2007.

El Niño, means "the child" in Spanish. The phenomenon is usually noticed around Christmas time ; El Nino is thus linked with "Christ" in the Pacific Ocean off the west coast of South America. La Niña means "the little girl". Their combined effect on climate is profound in the southern hemisphere. These climatic impacts were first described by Sir Gilbert Thomas walker in 1923. Walker Circulation, named after him, is prominent of the Pacific ENSO phenomenon.

In the Pacific, during major warm events, the warming of El Niño spreads, westwards, over much of the tropical regions of the Pacific Ocean and Indian oceans. ELNINO's impact varies with SO intensity. ENSO events in the Atlantic Ocean lag behind those in the Pacific by 12 to 18 months.

ENSO mostly affects developing countries in South America, Africa and South East Asia. These economies are largely dependent upon their agricultural and fishery sectors, whose source of food supply, employment, and foreign exchange are impacted. The onset of ENSO events in the three oceans do have global socio-economic impacts, adversely in general. While ENSO is a global and natural part of the Earth's climate, whether its intensity or frequency may change as a result of Global Warming, is a part of research presently..

El Niño and La Niña

El Niño and La Niña are officially defined as sustained sea surface temperature anomalies of magnitude greater than 0.5°C across the central tropical Pacific Ocean. When the condition is met for a period of less than five months, it is classified as El Niño or La Niña conditions; if the anomaly persists for five months or longer, it is classified as an El

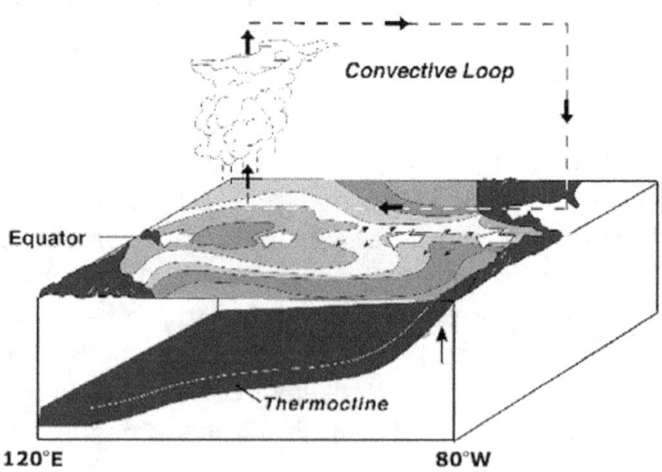

Fig. 22.1, Normal pacific pattern. Equatorial winds gather warm water pool toward west. Cold water up wells along South American coast. (NOAA / PMEL / TAO) [coloured version of the fig is on page 239]

Steps Towards A Green World

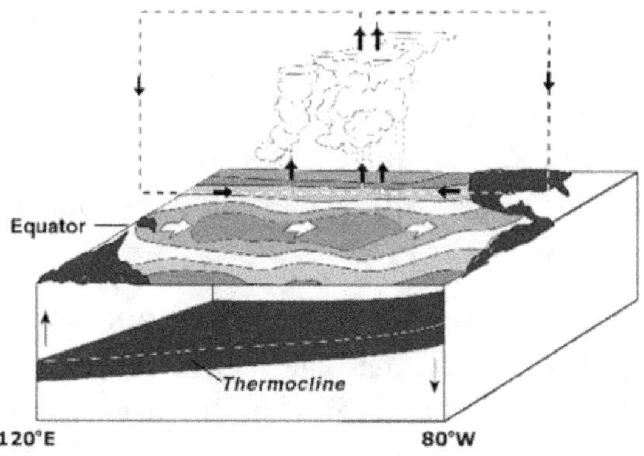

Fig. 22.2, El Niño conditions. Warm water pool approaches South American coast. Absence of cold upwelling increases warming.
[coloured version of the fig is on page 239]

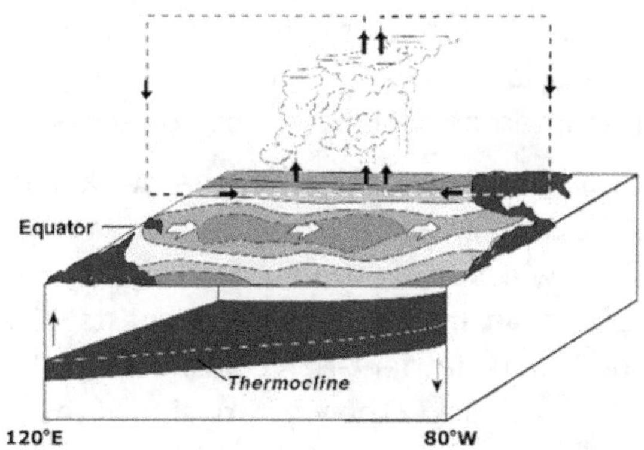

Fig.22.3, La Niña conditions. Warm water is further west than usual.
[coloured version of the fig is on page 240]

Niño or La Niña episode. Historically, it has occurred at irregular intervals of 2-7 years and has usually lasted one or two years.

The indicative signs of an El Niño are:

1. Rise in pressure of air, in the Indian Ocean, Indonesia, and Australia.

2. Fall in air pressure over Tahiti and the rest of the central and eastern Pacific Ocean.

3. Trade winds in the south Pacific weaken or head east.

4. Warm air rises near Peru, causing rain in the deserts there.

5. Warm water spreads from the west Pacific and the Indian Ocean to the east Pacific. It takes the rain with it, causing rainfall in normally dry areas and extensive drought in eastern areas.

El Niño's warm current of nutrient-poor tropical water, heated by its eastward passage in the Equatorial Current, replaces the cold, nutrient-rich surface water of the Humboldt Current, also known as the Peru Current, which abounds in food fish. In most years, the warming lasts only a few weeks or a month, after which the weather patterns return to normal and fishing improves. However, when El Niño conditions last for many months, more extensive ocean warming occurs and its economic impact to local fishing for an international market can be serious.

During non-El Niño conditions, the Walker circulation is seen at the surface as easterly trade winds which move water and air warmed by the sun towards the west. This also creates ocean upwelling off the coasts of Peru and Ecuador and brings nutrient-rich cold water to the surface, increasing fishing stocks. The western side of the equatorial Pacific is characterised by warm, wet low pressure weather as the collected moisture is dumped in the form of typhoons and thunderstorms. The ocean is some 60 cm higher in the western Pacific as the result of this motion.

Regional impacts of cold ENSO episodes (La Niña)

El Niño/Southern Oscillation (ENSO) - A shift in the normal relationship between the atmosphere and ocean in the tropical Pacific Ocean. Normally, strong winds (called 'trade winds' because they aid sailing ships transporting goods) blow to the west in the Pacific, moving warmer surface water away from North and South America. Simultaneously, cold water from the ocean depths rises to the surface off the west coast of South America. This upwelling brings nutrients to the surface, supporting fisheries and ecosystems in the area. In an El Niño event, these trade winds die down, causing warmer surface water to accumulate off western North and South America. This leads to increased rainfall, storm activity, and flooding in the Americas (especially the southwestern United States and Peru) and drought conditions in Australia and other areas in the western Pacific and the Indian Ocean. Fisheries on the west coasts of North and South America are also seriously affected.

Wider effects of El Niño conditions

Because El Niño's warm pool feeds thunderstorms above, it creates increased rainfall across the east-central and eastern Pacific Ocean.

The effects of El Niño in South America are direct and stronger than in North America. An El Niño is associated with warm and very wet summers (December-February) along the coasts of northern Peru and Ecuador, causing major flooding whenever the event is strong or extreme. The effects during the months of February, March and April may become critical. Southern Brazil and northern Argentina also experience wetter than normal conditions but mainly during the spring and early summer. Central Chile receives a mild winter with large rainfall, and the Peruvian- Bolivian-Altiplano is sometimes exposed to unusual winter snowfall events. Drier and hotter weather occurs in parts of the Amazon River Basin, Colombia and Central America.

Direct effects of El Niño resulting in drier conditions occur in parts of Southeast Asia and Northern Australia, increasing bush fires and worsening haze and decreasing air quality dramatically. Drier than normal conditions are also generally observed in Queensland, inland Victoria, inland New South Wales and eastern Tasmania from June to August.

West of the Antarctic Peninsula, the Ross, Bellingshausen, and Amundsen Sea sectors have more sea ice during El Niño. The latter two and the Weddell Sea also become warmer and have higher atmospheric pressure.

In North America, typically, winters are warmer than normal in the upper Midwest states, the Northeast, and Canada, while central and southern California, northwest Mexico and the southwestern US, are wetter and cooler than normal. Summer is wetter in the intermountain regions of the US The Pacific Northwest states, on the other hand, tend to experience dry but foggy winters and warm, sunny and precocious springs during an El Niño. During a La Niña, by contrast, the Midwestern US tends to be drier than normal. El Niño is associated with decreased hurricane activity in the Atlantic, especially south of 25° N; this reduction is largely due to stronger wind sheartropics.

Finally, East Africa, including Kenya, Tanzania and the White Nile basin experiences, in the long rains from March to May, wetter than normal conditions. There also are drier than normal conditions from December to February in south-central Africa, mainly in Zambia, Zimbabwe, Mozambique and Botswana.

Western Hemisphere Warm Pool

Study of climate records has found that about half of the summers after an El Niño have unusual warming in the Western Hemisphere Warm Pool (WHWP). This affects weather in the area and seems to be related to the North Atlantic Oscillation.

Atlantic effect

An effect similar to El Niño sometimes takes place in the Atlantic Ocean, where water along equatorial Africa's Gulf of Guinea becomes warmer and eastern Brazil becomes cooler and drier. This may be related to El Niño Walker circulation changes over South America.

Cases of double El Niño events have been linked to severe famines related to the extended failure of monsoon rains, as in the book Late Victorian Holocausts.

Non-climate effects

Along the west coast of South America, El Niño reduces the upwelling of cold, nutrient-rich water that sustains large fish populations, which in turn sustain abundant sea birds, whose droppings support the fertiliser industry.

East Pacific fishing

The local fishing industry along the affected coastline can suffer during long-lasting El Niño events. The world's largest fishery collapsed due to over-fishing during the 1972 El Niño Peruvian anchoveta reduction. During the 1982-83 event, jack mackerel and anchoveta populations were reduced, scallops increased in warmer water, but hake(fish) followed cooler water down the continental slope, while shrimp and sardines moved southward so some catches decreased while others increased. Horse mackerel have increased in the region during warm events.

Shifting locations and types of fish due to changing conditions provide challenges for fishing industries. Peruvian sardines have moved during El Niño events to Chilean areas. Other conditions provide further complications, such as the government of Chile in 1991 creating restrictions on the fishing areas for self-employed fishermen and industrial fleets.

The ENSO variability may contribute to the great success of small fast-growing species along the Peruvian coast, as periods of low population removes predators in the area. Similar effects benefit migratory birds which travel each spring from predator-rich tropical areas to distant winter-stressed nesting areas. There is some evidence that El Niño activity is correlated with incidence of red tides off of the Pacific coast of California.

It has been postulated that a strong El Niño led to the demise of the Moche and other pre-Columbian Peruvian cultures.

A recent study of El Niño patterns suggests that the French Revolution was caused in part by the poor crop yields of 1788-89 in Europe, resulting from an unusually strong El-Niño effect between 1789-93.

History of the phenomenon

ENSO conditions seem to have occurred at every two to seven years for at least the past 300 years, but most of them have been weak.

Major ENSO events have occurred in the years 1790-93, 1828, 1876-78, 1891, 1925-26, 1982-83, and 1997-98.

Recent El Niños have occurred in 1986-1987, 1991-1992, 1993, 1994, 1997-1998, 2002-2003, and 2006-2007, 2015.

The El Niño of 1997 - 1998 was particularly strong and brought the phenomenon to worldwide attention, while the period from 1990-1994 was unusual in that El Niños have rarely occurred in such rapid succession (but were generally weak). There is some debate as to whether global warming increases the intensity and/or frequency of El Niño episodes. (see also the ENSO and Global Warming sect.

23
PALEOCLIMATOLOGICAL ANALYSIS
(Climate Forming Over Millennia)

Paleoclimatology

It is the study of climate from the Big Bang onwards, millions of years, before our ancestors realised the importance of weathers and climates. Paleo-climatologists employ, primarily the following methodologies to formulate their theories and conclusions. These studies are reserved for specialists:

(i) Glaciers and ice domes, particularly recent findings of ice caps of Greenland and Antarctica reveal (a) pollen count indicates the total amount of plant growth of that year, (b) thickness of the layer gives precipitation quantity of that year, (c) trapped air in the ice layer helps determining the air composition of that period (d) warm and cold periods have different precipitations of heavier isotopes of hydrogen and oxygen, reflect cycles of average temperatures at ocean surfaces. Various cold and warm cycles have come to light. (ii) Rings from living trees can help Dendroclimatologists to estimate rainfall and temperature of recent times back to a few million years,(iii) Sedimentary rock layers with visible changes can give major changes in climate from prehistoric years, (iv) palynology, the science of pollens, can draw conclusions of climates from sediment layers of vegetation and animals' remnants at the bottom of oceans and lakes.

The Quaternary sub-era includes the current climate. There has been a cycle of ice ages for the past 2.2-2.1 million

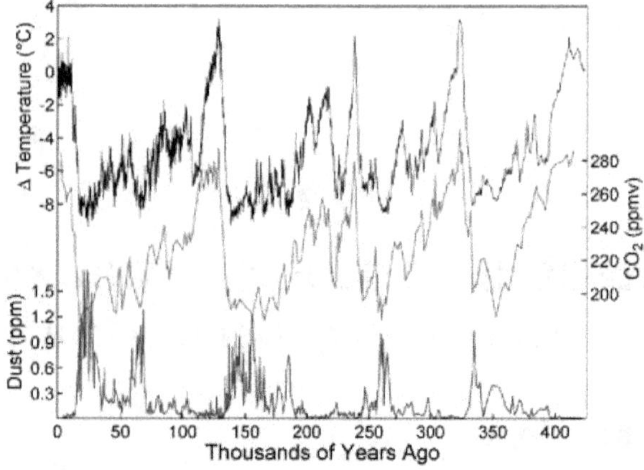

Fig. 23.1, Ice core data for the past 400,000 years. Note length of glacial cycles averages ~100,000 years. Blue curve is temperature, green curve is CO_2 and red curve is windblown glacial dust (loess). Today's date is on the left side of the graph.

years (starting before the Quaternary in the late Neogene Period).

Note in the graphic on the right the strong 120,000 year periodicity of the cycles, and the striking asymmetry of the curves. This asymmetry is believed to result from complex interactions of feedback mechanisms. It has been observed that ice ages deepen by progressive steps, but the recovery to interglacial conditions occurs in one big step.

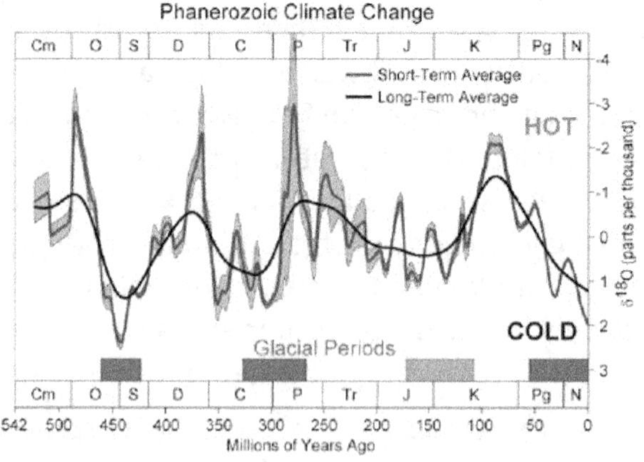

Fig. 23.2, 500 million years of climate change

Steps Towards A Green World

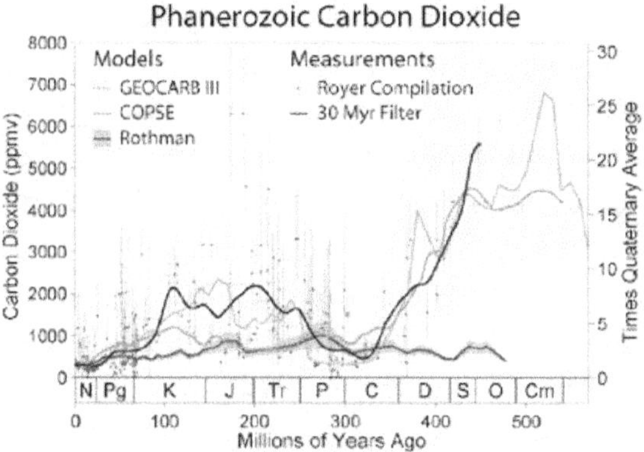

Fig.23.3, 500 million years of changes in carbon dioxide concentrations [coloured version of the fig is on page 240]

Controlling Factors of Climate

Geologically, short-term (<120,000 year) temperatures are believed to be driven by orbital factors (see Milankovitch cycles). The arrangements of land masses on the Earth's surface are believed to reinforce these orbital forcing effects.

Continental drift obviously affects the thermohaline circulation, which transfers heat between the equatorial regions and the poles, as does the extent of polar ice coverage.

The timing of ice ages throughout geologic history is in part controlled by the position of the continental plates on the surface of the Earth. When landmasses are concentrated near the polar regions, there is an increased chance for snow and ice to accumulate. Small changes in solar energy can tip the balance between summers, in which the winter snow mass completely melts and summers, in which the winter snow persists until the following winter. See the web site Paleomap Project for images of the polar landmass distributions through time.

Comparisons of plate tectonic continent reconstructions and paleoclimatic studies show that the Milankovitch cycles have the greatest effect during geologic eras when landmasses have been concentrated in polar regions, as is the case today. Today, Greenland, Antarctica, and the northern portions of Europe, Asia, and North America are situated such that a minor change in solar energy will tip the balance between year-round snow/ice preservation and complete summer melting. The presence of snow and ice is a well-understood positive feedback mechanism for climate. The Earth today is considered to be prone to ice age glaciations.

Another proposed factor in long term temperature change is the Uplift-Weathering Hypothesis, first put forward by T.C. Chamberlin in 1899 and later independently proposed in 1988 by Maureen Raymo and colleagues, where up thrusting mountain ranges expose minerals to weathering resulting in their chemical conversion to carbonates thereby removing CO_2 from the atmosphere and cooling the earth. Others have proposed similar effects due to changes in average water table levels and consequent changes in sub-surface biological activity and PH levels.

Over the very long term the energy output of the Sun has gradually increased, on the order of 5% per billion (109) years, and will continue to do so until it reaches the end of its current phase of stellar evolution.

Atmosphere's Three Eras

Our planet's modern atmosphere is called "third atmosphere". Dating back from 250000 million years ago it has been reasonably constructed by scientists. Beyond this era, prior to 1000,000 million years, the climatic conditions remain a mystery and are subject of intense research. Paleoclimatologists have tried to map, two more distinct atmospheric eras, different in chemical compositions, possibly drawing inferences from other sciences.

(i) The original atmosphere the "first atmosphere", immediately following the Big Bang, was primarily helium and hydrogen. Heat from the still-molten crust, and the sun, plus probably, enhanced solar wind, cooled off this atmosphere. It is believed that about 4.4 billion years ago, the fluidic cosmos was converted into a solid crust, at least in upper parts. But frequently erupting volcanoes released steam, carbon dioxide, and ammonia, meaning major atmospheric changes. This was termed the "second atmosphere", which was primarily carbon dioxide and water vapour, with some nitrogen but virtually no oxygen till 1.7 billion years back. But the atmosphere contained huge quantities of gases; approximately 100 times as much gas as the current atmosphere.

(ii) Continuous cooling pushed much of the carbon dioxide into seas and precipitated out as carbonates, leaving in the atmosphere primarily nitrogen and carbon dioxide and up to as much as 40 % hydrogen as established by research at the University of Waterloo, Canada, and University of Colorado, USA, in 2005. Fortunately, the greenhouse effect due to high levels of carbon dioxide and methane, kept the Earth from freezing. In fact temperatures were probably very high, over 70°C (158 degrees F), until some 2.7 billion years ago.

(iii) Fossil evidence indicates that bacteria shaped like cyanobacteria existed approximately 3.3 billion years ago and were the first oxygen-producing evolving phototropic organisms. They were responsible for the initial conversion of the earth's atmosphere from a state without oxygen (anoxic) to a state with oxygen (oxic) during the period 2.7 to 2.2 billion years ago. That is, the first ever oxygenic photosynthesis took place converting carbon dioxide into oxygen, playing a major role in oxygenating the atmosphere.

Photosynthesising of plants, later convert more carbon dioxide into oxygen. However, over time, the excess carbon was locked in fossil fuels, sedimentary rocks (notably limestone), and animal shells. Oxygen in turn reacted with ammonia to release nitrogen. In addition, bacteria also converted ammonia into nitrogen. But most of the nitrogen currently present in the atmosphere results from sunlight-powered photolysis of ammonia released steadily over the aeons from volcanoes.

As more plants appeared, the levels of oxygen increased significantly, while carbon dioxide levels dropped. At first the oxygen converted metals into oxides (as iron), but eventually oxygen accumulated in the atmosphere, resulting in mass extinctions and further evolution. With the appearance of an ozone layer (ozone is 3 atoms of oxygen) life forms were better protected from ultraviolet radiation. This oxygen-nitrogen atmosphere is the "third atmosphere". 200 - 250 million years ago, up to 35 per cent of the atmosphere was oxygen (bubbles of ancient atmosphere were found in anamber).

The composition of modern atmosphere is enforced by oceanic blue-green algae as well as geological processes. O_2 does not remain naturally free in an atmosphere, but tends to be consumed (by inorganic chemical reactions, as well as by animals, bacteria, and even land plants at night), while CO_2 tends to be produced by respiration and decomposition and oxidation of organic matter. Oxygen would vanish within a few million years due to chemical reactions and CO_2 dissolves easily in water and would be gone in millennia if not replaced. Both are maintained by biological productivity and geological forces seemingly working hand-in-hand to maintain reasonably steady levels over millions of years (see Gaia theory).

Atmospheric gases scatter blue light more than other wavelengths, giving the Earth a blue halo when seen from space.

Glaciers

This image shows the termini of the glaciers in the Bhutan-Himalaya. Glacial lakes have been rapidly forming on the surface of the debris-covered glaciers in this region during the last few decades. USGS researchers have found a

Glaciers

strong correlation between increasing temperatures and glacial retreat in this region.

Glacial motion is the motion of glaciers, which can be likened to rivers of ice. It has played an important role in sculpting many landscapes. Most lakes in the world occupy basins scoured out by glaciers. Glacial motion can be fast (up to 30 m/day, observed on Jakobshavns Isbrae in Greenland) or slow (0.5 m/day on small glaciers or in the center of ice sheets).

Glacier motion occurs from two processes, both driven by gravity: basal sliding and internal deformation. In the case of basal sliding, the entire glacier slides over its bed. This type of motion is enhanced if the bed is soft sediment, if the glacier bed is thawed and if melt-water is prevalent. A glacier that is frozen up to its bed does not experience basal sliding. Internal deformation, the second process, occurs when the weight of the ice causes the deformation of ice crystals. This takes place most readily near the glacier bed, where pressures are highest. There are glaciers that primarily move via sliding, and others that move almost entirely through deformation.

If a glacier's terminus moves forward faster than it melts, the net result is advance. Glacier retreat occurs when more material ablates from the terminus than is replenished by flow into that region.

Glaciologists consider that trends in mass balance for glaciers are more fundamental than the advance or retreat of the termini of individual glaciers. In the years since 1960, there has been a striking decline in the overall volume of glaciers worldwide. This decline is correlated with global warming. As a glacier thins, due to the loss of mass it will slow down and crevassing will decrease.

At some point, if an Alpine glacier becomes too thin it will stop moving. This will result in the end of any basal erosion. The stream issuing from the glacier will then become clearer as glacial flour diminishes.

During the Pleistocene (the last ice age), huge sheets of ice called continental glaciers advanced over much of the earth. The movement of these continental glaciers created many now-familiar glacial landforms. As the glaciers were expanded, due to their accumulating weight of snow and ice, they crushed and redistributed surface rocks, creating erosional landforms such as striations, cirques, and hanging valleys.

Later, when the glaciers retreated leaving behind their freight of crushed rock and sand, depositional landforms were created, such as moraines, eskers, drumlins, and kames. The stone walls found in New England (northeastern United States) contain many glacial erratics, rocks that were dragged by a glacier many miles from their bedrock origin.

Lakes and ponds can also be caused by glacial movement. Kettle lakes form when a retreating glacier leaves behind an underground chunk of ice. Moraine-dammed lakes occur when a stream (or snow runoff) is dammed by glacial till.

Studying glacial motion and the landforms that result requires tools from many different disciplines: physical geography, climatology, and geology are among the areas sometime grouped together and called earth science.

Percentage of advancing glaciers in the Alps in the last 80 years.

Atmospheric Impacts

(i) The past decade was the warmest in more than 4 centuries. The rate of warming is unprecedented 10000 years. IPCC predicts temperature rise between 1.8 to 4 °C by the end of the century, (ii) Sea level today is 15 cm higher than a 100 years ago, will go up by another 20 cm by 2030 due to glacial melting and thermal expansion of oceans. Melting of Greenland's ice cap will add 6 meters to sea-level, (iii) It has been conclusively proven that the ozone layer is thinning through out the world. A 20% increase in the air's ozone level is expected to cut crop yields by 20 %, affecting India, China, Europe, US and Middle East by 2050.

Impact on Humans

(i) Rise in temperatures will put 400 million people at risk of hunger, 1-3 billion at risk of water stress and reduce crop output between 20 to 400 million tons, (ii) WHO notes that global warming increases risk of cholera, malaria, and infectious diseases; these diseases today lead to 150000 deaths and 5 million diseases, (iii) 1% loss of ozone leads to 3-5% increase in skin cancer. Droughts, floods and rising seas will force millions to seek new homes. Today there are 30 million environmental refugees; the number will go up to 50 million by 2010.

Impact on Biosphere

(i) Flowering and leaf flushing occur 2.3 days earlier every year, affecting movement of species every year. By 2050 plains may become deserts and will alter the ecology of forests, leading to extinction of one fourth (about a million) of species of plants and animals, (ii) Polar bears have dropped in numbers and weight in Arctic. Emperor penguins have lesser breeding pairs-300 to just 9- in the western Antartic Peninsula. Between 100-200 other cold dependent species are in deeper trouble.

Impact on Weather Patterns

(i) It is likely that cyclones, typhoons and hurricanes in the future will have higher peak wind speeds and more heavy precipitation on account of sea-surface warming. 30% of coastal wet lands will be lost by 2050; by then more than a billion people will be water starved (ii) Snowfall and rainfall patterns will change, leading to more draughts and floods. Higher temperatures mean more demand of water, in houses and for crops. Water quality may alter owing to increase in the incidence of algal blooms.

World Glacier Monitoring Service

The World Glacier Monitoring Service (WGMS) was started in 1986, combining the two former services PSFG (Permanent Service on Fluctuations of Glaciers) and TTS/WGI (Temporal Technical Secretary/World Glacier Inventory).

WGMS "collects standardised observations on changes in mass, volume, area and length of glaciers with time (glacier fluctuations), as well as statistical information on the distribution of perennial surface ice in space (glacier inventories). Such glacier fluctuation and inventory data are high priority key variables in climate system monitoring; they form a basis for hydrological modelling with respect to possible effects of atmospheric warming, and provide fundamental information in glaciology, glacial geomorphology and quaternary geology."

Steps Towards A Green World

24

EFFECT ON ENVIRONMENT
by Glaciers, Air, Sun's Radiations, Water and Soil

Truly Pristine forests, grasslands, and wild fauna and flora, are few and far between. Humans have occupied virtually all the lands and seas of this globe. Research, by almost all major countries, is on for spreading out to the Arctic and Antartic continents. The world powers are in a race to exploring, (controlling) habitability of space in the cosmos. Modern day environmentalists (political, social, and philosophical idealists) limit advocating various actions and policies in the interest of protecting whatever remains natural, or restoring or expanding the role of nature in the existing environment. Hence the understanding of natural environment by present day ecology and earth watchers differs from the one adopted in this chapter.

The natural environment, or simply the term environment, comprises of the atmosphere above the earth besides all living and non-living things that occur naturally on earth or some part of it, (e.g. the natural parks in India, the USA, Australia and Africa), prior to industrialisation. The key components of this term are:

1. Complete landscape units/ geographical areas that have survived and still function as natural systems where human intervention is negligible. The natural phenomena occur unhindered within their boundaries on plants, living species, rocks and soil; generally termed

wilderness. A geographical area, limited, can be regarded as a natural environment, if it is maintained under certain defined conditions. These conditions vary from country to country, based on developments, since inception, due to natural phenomena on particular geographical location.

2. Natural resources and phenomena that exist universally, without boundaries, such as sun, air, water, soil and the climate resulting from their influences.

3. Orbital spinning of the planet, round its own axis and around the sun, is a part of the natural environment (for our purposes).

The virgin natural environment lasted for millions of years. Its dilution, noticeably, commenced after the onset of industrialisation in the 18th. Century. The processes described below speak of the bygone millennia.

Glaciers

Glaciers are recognised as one of the most sensitive indicators of climate change. Glaciers grow and collapse, both contributing to natural structural changes and greatly amplifying externally-forced changes. These phenomena led to formation of landmass, which drifted to become continents. To elaborate, the most significant climate processes of the last several million years are the glacial and interglacial cycles of the ice age. Though shaped by orbital variations, the internal responses involving continental ice sheets and 130 million years sea-level change certainly played a key role in deciding shapes of climates of most regions. The movements, very slow though, rubbed against tectonic rocks beneath them and pulverised their upper surfaces. This process left beds of dust /mineral particles where from the glaciers receded, which were flown into the atmosphere by high winds. Some other types of changes, including *Heinrich events, Dansgaard–Oeschger events

Steps Towards A Green World

and the Younger Dryas have shown the potential for glacial variations to influence climate even in the absence of specific orbital changes.

Since the last century, however, glaciers have not received enough ice during the winters; overall more ice was lost during the summer months, diminishing stored ice, called glacier retreat. No surprise, therefore, that the global warming presently is focusing on the melting of islands of ice, around north of Canada, in the Arctic seas, on slopes of the Alps and Himalayas, and the Polar Regions of Antarctica.

Sun's Radiations

The range of electromagnetic rays, emanating from the sun is virtually infinite. However, the Sun's radiations primarily can be divided in to heat causing and light giving ones. The energy output of the sun, which is converted to heat raises temperature of the ground, air in the atmosphere and water surfaces. Heated currents of air and water vapour lead to weathers of myriad kinds as explained later. Suffice it to say that the heat giving energy of the sun plays the most significant role, directly and indirectly, in climate forming. On the other hand, the rays resulting in light cause photosynthesis of plants, a phenomenon which absorbs CO_2 (converts it into carbonates) and releases oxygen into the air. CO_2 is becoming the significant constituent in greenhouse gases and influences climates effectively. By reversing the process of photosynthesis, removing CO_2 from the environment, may help, in the future, in slowing global warming. The rays of infra red, ultraviolet, gamma and x- ray frequencies, also affect climate and humans in many ways.

A variety of forms of solar energy variation, including the 11-year solar cycle, (sun is getting brighter with higher energy output) and longer-term modulations, are taking place. However, the 11-year sunspot cycle does not manifest itself clearly in the climatological data. Solar intensity variations are considered to have been influential in triggering the little ice age, and for some of the warming observed from year 1900 to 1950. The cyclical nature of the sun's energy output is not yet fully understood. However; continuing its main sequence, this slow change or evolution affects the earth's atmosphere as also the global climates.

Solar variations

Variations in solar activity during the last several centuries based on observations of sunspots and beryllium isotopes.

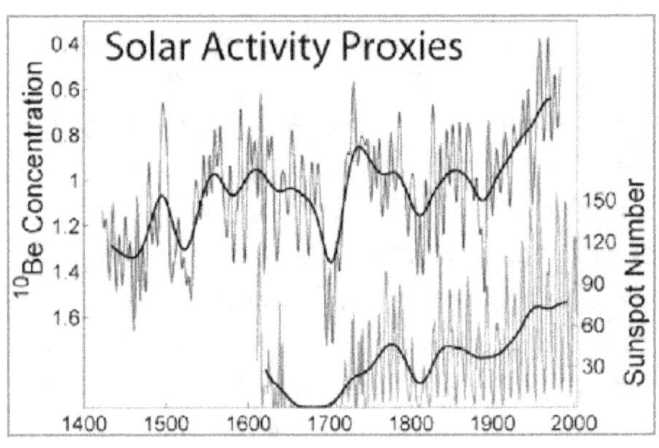

Fig. 24.1, Solar activity proxies

Air

Air is a combination of oxygen and nitrogen and it surrounds the earth up to a height of 1,000km. In prehistoric` eras, firstly, rocks were broken up by onslaught of storms caused by changing temperature. Warm and hot ground surfaces force the air upwards. The air from cooler parts moves in, to the higher temperature areas. Thus, wind (moving air) movements of varying speeds, rub rocks and stones, breaking them up, bit by bit by friction, reducing them to small particles over a long time. Sand and particle dusts, in existence from earlier periods, swirl and rise with the wind into higher atmosphere. Dust storms in deserts,

Steps Towards A Green World

like those from Rajasthan over Delhi and Haryana, are an accepted fact.

The sun heats water bodies like surfaces of rivers, lakes and seas, as well. The evaporated water rises as vapour and gets mixed in the wind which also contains particles of dust. Cloud formation, lightning strokes and rain like phenomena take place in Troposphere, a height of about 15 km.

Nall (as cited in "References") called moisture ratio (mass or volume of air / mass or volume of moisture in that air) as one of "the most important climate variables for human comfort and designing buildings for that climate to obtain energy efficiency".

Since, many millenia sun's radiations have released huge quantities of oxygen while absorbing CO_2 given out by algae in oceans and water bodies and methane from buried coal, natural gas and oil deposits. When forests and wild plantations became abundant, the environment became highly oxidising from a 'chemically reducing era'. Till present times there were many cyclic periods when the quantity of CO_2 rose very high and altered climates on our planet. Though CO_2 is a small percentage of the atmospheric mass, its rise in recent times is causing anxiety, as its role in warming the globe is appreciable (Refer to Appendix 2 for more on CO_2).

Troposphere contains Ozone layer, from 15km to 35/50 km. Within this layer, the air gets thinner and ozone increases with the distance, though only a few parts in a million. Ozone blocks the harmful ultra violet rays. Hence, its depletion in recent years is causing concern, because more of the UV rays reach the earth (*In chapter 21, the 'Ozone Layer' is discussed comprehensively*).

Orbital Revolutions of the Earth

It is well known that the earth revolves round its axis as

well as the sun. Our days and seasons are closely related to these orbital movements. A change in sun's distance from the earth means higher or lower amounts of sun's radiations reaching the earth. These cause variations in temperature and absorption of electromagnetic rays released by the sun; considered by scientists, the driving factors underlying the glacial and interglacial cycles of the present ice age. Subtler variations leading to repeated advance and retreat of the Sahara desert in response to orbital precession are also known.

Some such orbital variations, known as Milankovitch cycles, have highly predictable consequences; as they are based on the mutual interactions of the Earth, its moon, and the other planets.

Plate Tectonics

On the longest time scales, plate tectonics have repositioned continents, shaped oceans, built and tore down mountains and generally, served to define the stage upon which climate exists. Approximately 3 million years ago, the North and South American plates collided to form the Isthmus of Panama and shut off direct mixing between the Atlantic and Pacific Oceans. Since then there has been intensification of the present ice age. Many similar climatic changes would have occurred.

Volcanic Eruptions

The eruption of Mount Pinatubo in 1991 is barely visible on the global temperature profile. However, several eruptions of that kind per century can cause cooling for a period of a few years. Much stronger volcanic eruptions occur only a few times every hundred million years. They, however, can reshape climate for millions of years and cause mass extinctions. Initially, scientists thought that the dust emitted into the atmosphere from large volcanic eruptions was responsible for the cooling by partially blocking the transmission of solar radiation to the Earth's surface. However, measurements indicate that most of the dust thrown in the atmosphere returns to the Earth's surface within six months.

Volcanos are also part of the extended carbon cycle. Over very long (geological) time periods, they release carbon dioxide from the earth's interior, counteracting the uptake by sedimentary rocks and other geological carbon sinks. However, this contribution is insignificant as compared to the current anthropogenic emissons. The US Geological Survey estimates that human activities generate 150 times the amount of carbon dioxide emitted by volcanoes.

Water

The presence of water made habitation possible on our planet. Flowing waters in rivers constantly cut their beds and banks. They are the principal sources of sand and mineral particles. Flooded rivers, especially the rain streams, allow rapid drainage of heavy rains-lay of new soil over existing landscapes. Some tracts become more fertile and others turn barren. Geographic variations in vegetation change environments and impact on climates.

Of late, waters of almost all the world rivers are depleting. Reservoirs and dams built to generate electricity regulate outflows. Factories draw huge quantities of water, directly or otherwise and throw out polluted waters. These effluents from factories are killing fishes and aqua based species. Forests on their banks are getting eroded. The water caused pollution causes alterations in environments.

A positive outcome and of immediate importance are the algae and bacteria which are born in water bearing soil, ponds, lakes and seas and in thick forests. Leaving aside their role in the paleoclimatology and historic climates, presently, most of these biological miniscule cells are supporting new plants and products which are providing new hopes of survival on the planet. Using micro-biological techniques, huge quantities of bio-masses of a large variety can be ushered in. On the one hand, these can become sources of renewable energy. They, on the other hand, are leading to 'zero waste agriculture'. The byproducts, waste, of crops are converted into manure and pesticides and weed killers, replacing chemical fertilisers and insecticides. The organic vegetables and fruits are healthier than fertiliser and pesticide supported crops. It seems that the genic agents of pre-historic times will become our saviours in the future.

The Role of Water Vapour

Intergovernmental Panel on Climate Change (IPCC) IPCC Third Assessment Report discusses the impact of water vapour in detail.

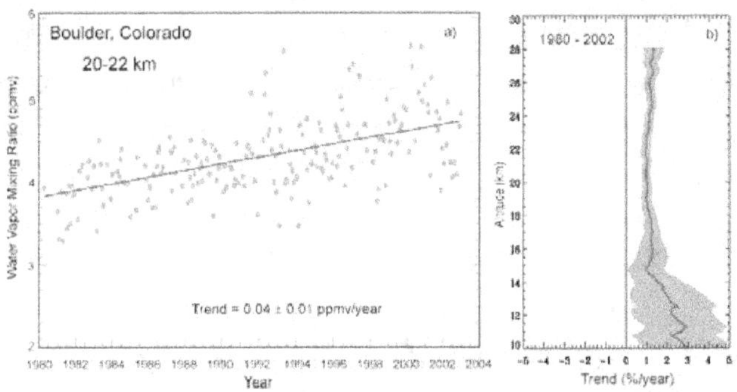

Fig. 24.2, Increasing water vapour at Boulder, Colorado

Water vapour is a naturally occurring greenhouse gas and accounts for the largest percentage of the greenhouse effect, between 36% and 90%. Water vapour concentrations fluctuate regionally. However, human activity does not directly affect water vapour concentrations except at local scales (for example, near irrigated fields).

An increase in atmospheric temperature in turn leads to an increase in the water vapour content of the troposphere.

The increased water vapour in turn leads to an increase in the greenhouse effect and thus, a further increase in temperature; the increase in temperature leads to further increase in atmospheric water vapour; and the feedback cycle continues until equilibrium is reached. Thus, water vapour acts as a positive feedback to the climate forcing factors provided by human-released greenhouse gases. Changes in water vapour may also have indirect effects via cloud formation and rain. A discordant note is struck by Michael Mann who says, "It is extremely misleading, however, when scientists cite the role of water vapour as a greenhouse gas," because it can not be controlled by humans.

Global Scenario of the Temperature

Only since 1850, measurements of the temperature by thermometer have been systematically recorded. Satellites have been measuring the temperature of the troposphere since 1979. Balloon measurements begin to show an approximation of global coverage since the 1950's.

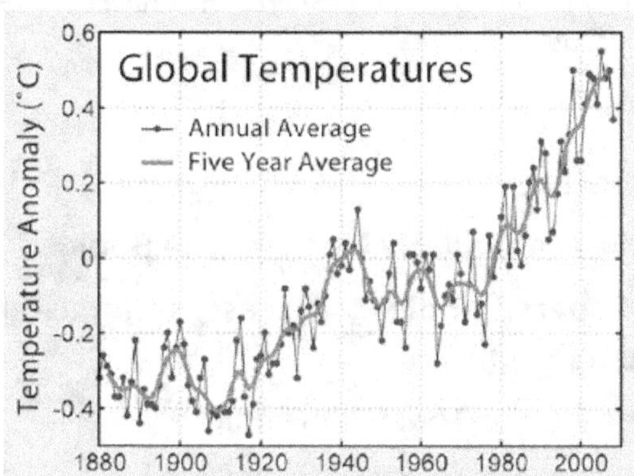

Fig. 24.3, The instrument recorded temperatures from 1850
[coloured version of the fig is on page 233]

Several groups have analyzed the satellite data to calculate temperature trends in the troposphere; RSS find

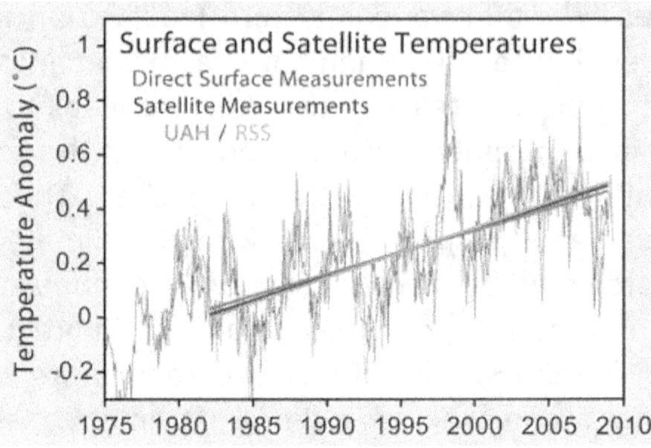

Fig. 24.4, Tropospheric temperature from 1975 (the satellite and balloon temperature records)

a trend of +0.208 °C/decade while UAH find +0.123 °C/decade; Fu et al find trends (up to the end of 2004) of +0.19 °C/decade when applied to the RSS data set; Vinnikov and Grody find +0.22°C to +0.26°C per decade (Oct. 03).

Both the graphs indicate abnormal rate of rise, of the temperature, from 1970 to 2000.

Note: Chapters 7 and 8 in Part II of the book are drawn from Wikipedia. These are based on experts researches. Therefore, no alteration is made.

Sources

1. **IPCC Reports**

2. **Wikipedia (particularly Part II of the Book)**

3. **News Papers: Indian Express, Hindustan Times, Hindustan (Hindi)**

4. **Magazines: India Today, Outlook**

Note: Suitably rephrased, made easier for readers

25

GLOSSARY

The author has paraphrased the following, in simple terms, to bring out the correct gist of the listed glossary. For the classic meanings, the author abides by descriptions of these terminologies as given in the text.

Anthropogenic: Polluting of atmosphere caused by "human' activities; e.g. Power Plants, Industries, Transportation, Agriculture etc. Mostly CO_2, methane and nitrous oxide are released.

CFCs: Chlorofluorocarbons are refrigerants which 'were' used in air conditioning systems and refrigerators. Their use was banned under Montreal treaty.

Climate is recognised, (by common people) by patterns over long periods, of variation in sunlight during days and nights, maximum and minimum temperatures, wind speeds, and rain/snow fall. Climate differs from weather, in that weather only describes the short-term (a day to a week) conditions of these variables in a given region. Climate is weather lasting over a few months. Climates generally depend upon longitude, latitude and height above sea level and nearness to water bodies. The mean values of each of the variables is recorded from year on year, and the pattern is established.

Meteorologists define climate by measurements, over long durations (30 years generally), temperature humidity, atmospheric pressure, wind, precipitation, atmospheric particle count and a number of other (may be 50 or more in all) factors in a given region.

Climate Change: Long term (5 to 10 years) 'perceptible variations,' in atmosphere, in particular, (a) of day and night temperatures, (b) frequent, unexpected rains causing floods and storms like cyclones, typhoons and hurricanes,, more frequent draughts, etc.

COPs: Conference of Parties, i.e. Meetings, are held annually; the first was held in 1995 and 21st in Paris. Parties means representatives of Nations who were a Party to UNFCCC. Generally, these representatives are of "Minister's Rank" (Indian Example), who can take decisions on behalf of their governments. The meetings discuss and decide on reports made by IPCC as well as on their respective government's suggestions.

Earth Summit: Meteorologists, scientists from physical. botanical, and biological sciences, representatives from IPCC, UN, WMO, UNEP, economists, NGOs and political and social scientists along with 172 representatives (116 Heads + 56 Delegates) of world nations, met in Rio, Brazil, in June 1992 to discuss, sustained Biological Diversity, Temperature Rise, and Deforestation. The Summit led to the formation of UNFCCC (formally ratified in 1994) and World Commissions on Bio-Diversity and Deforestation. Sadly, many of the initiatives at the Earth summit were passed over at Paris Summit in 2015. The proposition to formulate an 'international treaty' to rein in emission of greenhouse gases by human-related activities was generated; led to Kyoto Protocol in 1997.

Greenhouse Effect: Sun's radiations, particularly UV and infrared rays, get trapped in lower atmosphere, (7 to 10 Km above earth's surface) causing rise of temperature, which in turn release gases termed greenhouse gases.

Greenhouse Gases: Gets the name from glass covered greenhouses in nurseries in which seedlings of plants (trapping heat from sun's radiations) are prepared. The trapped heat and water vapour release a number of gases, mainly carbon dioxide, methane and nitrous oxide.

HCFCs: Hydro chlorofluorocarbons replaced CFCs as refrigerant. Their use has also been banned by Montreal Treaty.

IPCC: Intergovernmental Panel on Climate Change:- To all purposes it is the scientific and technical body of experts who prepare "reports" and "discussion papers", on behalf of UNFCCC, for COPs. The first Report appeared in 1991, followed by annual reports. Its Report of 2007 was seminal and the one of 2014 guided proceedings of the Paris Summit. All the Reports are, collection and summarised versions, of hundreds of researchers on sustained economic, environmental and social development of nations of the world.

Kyoto Protocol: An 'International Treaty' approved in 1997 by delegates of nations at a conference in Kyoto, Japan. The Protocol was ratified by required number of nations by 2005. It created Annexure I (rich and developed), and Annexure II (developing and poor) nations. Annexure I nations were set targets of reducing emission of greenhouse gases. Annexure II nations were advised to undertake measures of " Comparative Differential Responsibility" of emission mitigation and adaptation. Policies and processes of 'Carbon Trading", 'Transfer of Technology' and help with "Funds" were laid down. Though originally, the Kyoto Protocol was to end by 2012, it has been extended to 2020 when Paris Treaty will become effective.

Montreal Treaty: The first ever international treaty of its kind, signed in 1987, which banned, progressively, use of CFCs and HCFCs, refrigerants. The Treaty, ably documented, has been the most successful international treaty which helped stop depletion of the "Ozone Layer".

Ozone Layer: Ozone Gas, (an isotope of Oxygen with

its 3 atoms), is released, by photochemical reaction with refrigerants, CFCs or HFCls. The gas gets collected, and forms a layer, in the Trato-sphere, about 7-10 km above the Earth's surface. The layer blocks sun's infra red rays from reach the Earth and thus reduce the rise of temperature. The 'Depletion of the Ozone Layer is causing worry; the Earth's temperature rise may get aggravated.

Palaeoclimatological Analysis: Study and analyses of climates which got formed over millennia in the past.

Seasons are another form of recognising climate. by seasons. Seasons result from the yearly orbit of the Earth around the Sun and the tilt of the Earth's rotational axis relative to the plane of the orbit. The seasons are marked by weather, duration of sunlight and ecological behaviour.

A year is mostly divided in to four seasons; Winter, Spring, Hot, and Autumn. At the two poles, there are only two seasons; Winter (animals go in to hibernation and plants remain dormant) and Summer. The intensity of sunlight varies. Similarly two seasons happen in regions near the equator; rainy and dry or hot seasons. In some parts of the world, special "seasons" are loosely defined as hurricane season, tornado season, or a wildfire season.

Northern and Southern Hemispheres have opposite season; Winter in the Northern parts is Summer in the Southern Hemisphere and so on.

UNEP: The United Nations Environment Programme (UNEP), was set up in 1972, with headquarters near Nairobi, Kenya. The agency coordinates the UNITED Nation's environmentally sound policies and practices in Developing Countries. UNEP covers a wide range of issues, funding and implementing green economy projects, regarding the atmosphere, marine and terrestrial ecosystems, environmental governance and green economy. UNEP along with WMO, helped form IPCC in 1988. The Duo also help organise Earth Summit in 1992, when the idea of setting up UNFCCC was mooted.

The winner of the Miss Earth beauty pageant serves as the spokesperson of UNEP

UNFCCC: United Nations Framework Convention on Climate Change was formally set up in 1994. It is the United Nations General Assembly's approved body to frame policies and rules to control emission of Green House Gases. The "Objective" is to control rise of temperature to 2°C and ensure that GHG emissions of all countries reach their 'maximum level' between 2060-2080. And ensure that a sustained environment is possible; there is ecological balance. It has 196 members who nominate COPs.

Weather: In essence Weather is a short term (a day, a week or two) happenings of temperature, humidity, precipitation, and atmospheric conditions (virtually all factors/data that affects climates). Day to day recordings, are averaged over similar periods for a year. These very averages over a region, of area of decent magnitude get formulated into long term patterns which give identity to climates. The public is aware of daily weather forecasts; some are specific to agriculture, aviation, and maritime activities. Weather forecasts of rain/snowfall, draughts, months in advance, help governments plan remedial measures. However, the day to day weather forecasts generally don't come true; thus becoming a joke of parlance.

WMO: The World Meteorological Organisation (WMO) is the successor to the International Meteorological Organisation (IMO), founded in 1873. Established in 1950, WMO is the United Nation's official body for recording the state of all parameters affecting global weather and developing their long term pattern; defining climates. Besides, WMO records the interaction between the Earth's atmosphere with the oceans, the climate it produces and the resulting distribution of water resources. WMO, on 1 January 2013 had a membership of 191 Member States and Territories.

As weather, climate and the water cycle affect international regions, WMO promotes setting up of networks for exchanging information on meteorological, climatological, hydrological and geophysical observations. Additionally, it develops, along with other nations, standarisation of measuring instruments, processing, formalising and exchange of data. WMO's experts assist technology transfer, in training, setting up research projects, technology transfer and even collaboration between the National Meteorological and Hydrological Services of its Members.

WMO provides, publically, weather services to agriculture, aviation, shipping, the environment, water issues and the mitigation of the impacts of natural disasters. Its advance warnings of disasters, natural and human caused (such as accidents at chemical and nuclear plants, forest fires and overflow of volcanic ash) save lives and contain damage to properties.

———

ANNEX

Diagrams and Pictures in Colour

Some of the Black and White (B&W) diagrams and pictures in Part I and Part II of this book are reproduced in colour in this Annex, as they appear in the 'source' from which they are resourced. The colour version will help explain the text better. Underneath each B&W diagram, in the text, reference is given to the page no. of its corresponding colour version.

In these pages, cross-reference is given, underneath the colour version, to the page no. on which the corresponding B&W version appears in the text . For example, on page 233

Fig. 1.2 The frequency of floods has increased three-to four-fold (page 19) indicates that Fig. 1.2 B&W version appears on page 19.

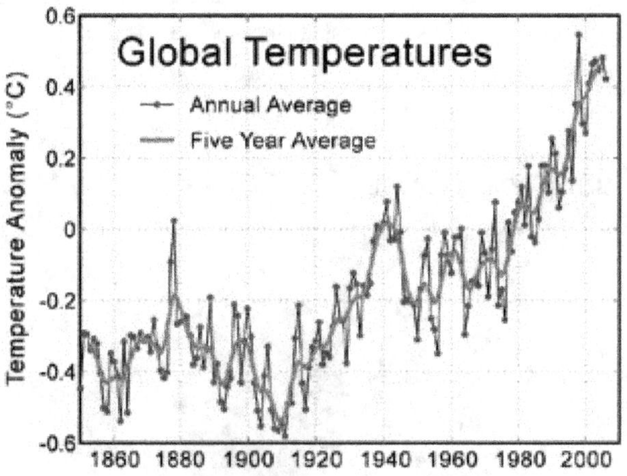

The instrument recorded temperatures from 1850

[Fig.1.1, (Page 12), Fig.17.1, (Page 160), Fig.24.3, (Page 225)]

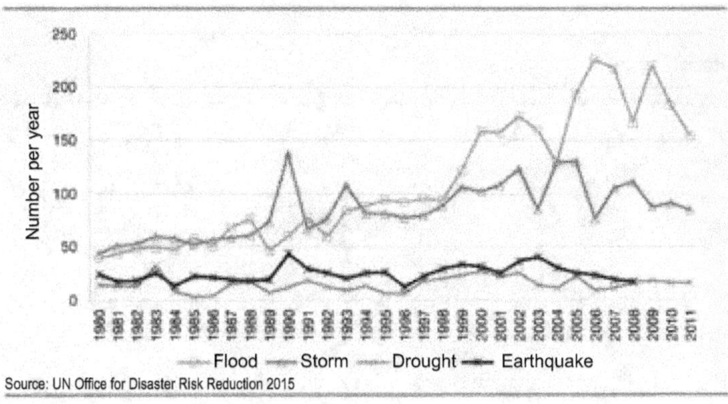

Fig.1.2, The frequency of floods has increased three-to four-fold (Page 19)

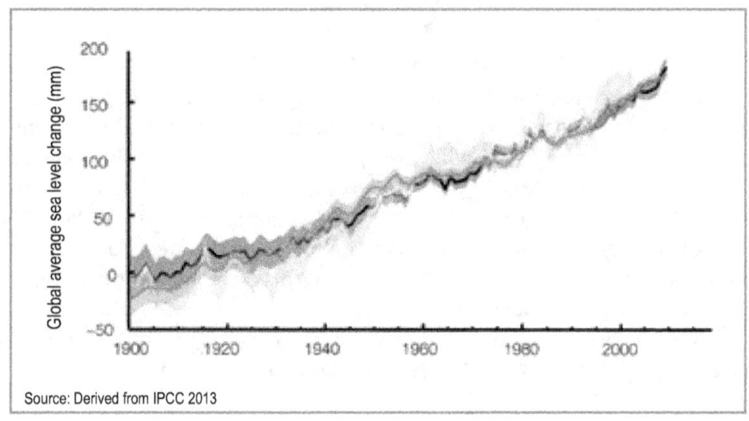

Fig.1.3, Sea levels has been rising since the late 1800s (Page 19)

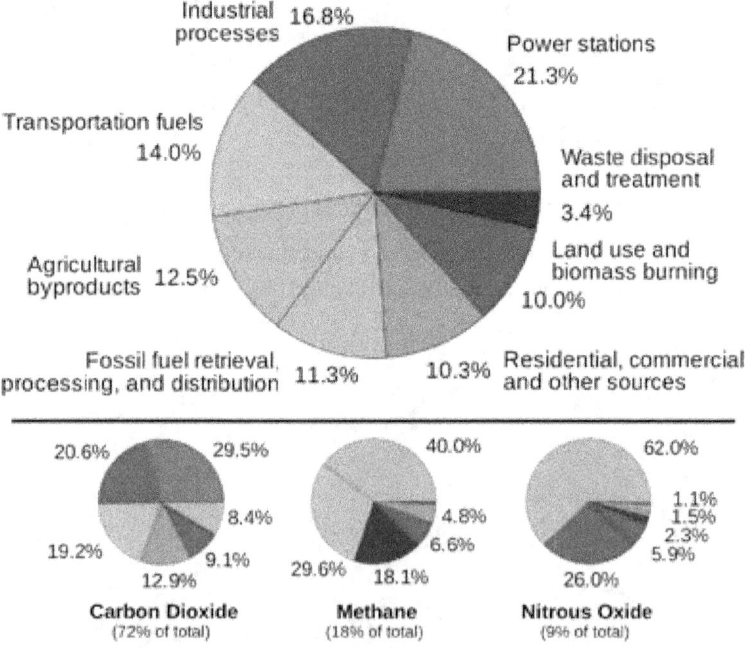

Fig. 2.3, Sector-wise annual generation and emission of greenhouse gases
(Page 37)

Steps Towards A Green World

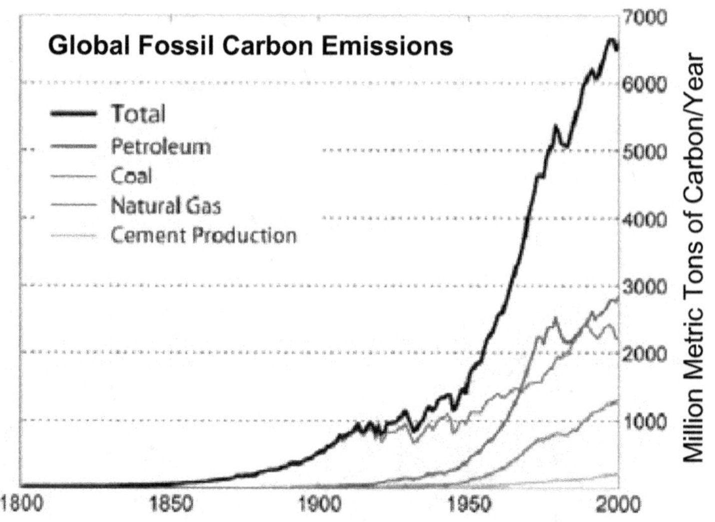

Fig. 2.4, Global carbon dioxide emissions 1751–2000 (Page 38)

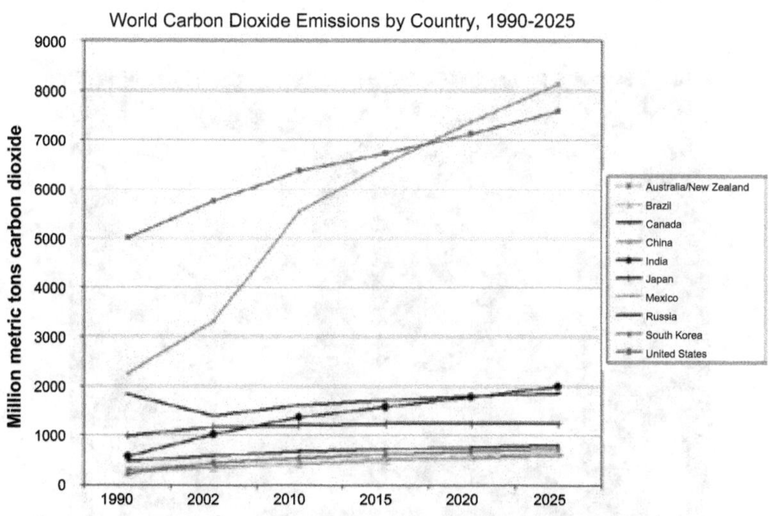

Fig. 2.5, Historical and projected CO_2 emissions country wise
(Source: Energy Information Administration) (Page 39)

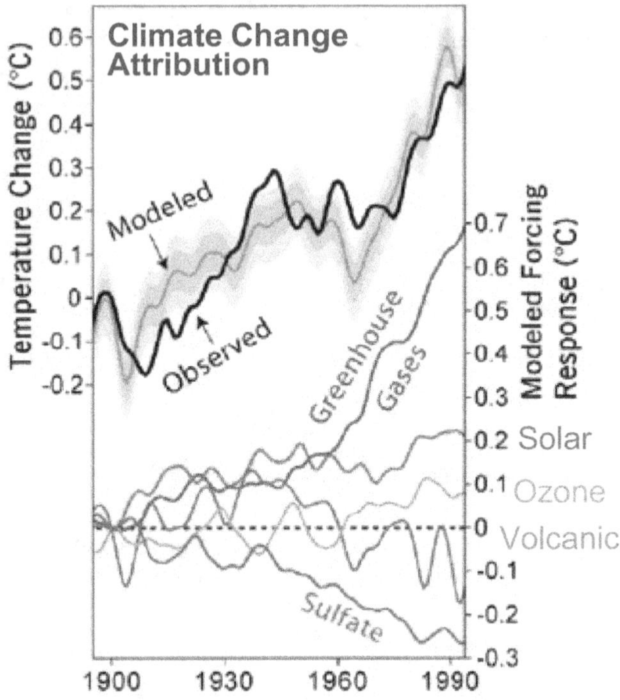

Fig. 2.6, Attribution of recent climate change (Page 39)

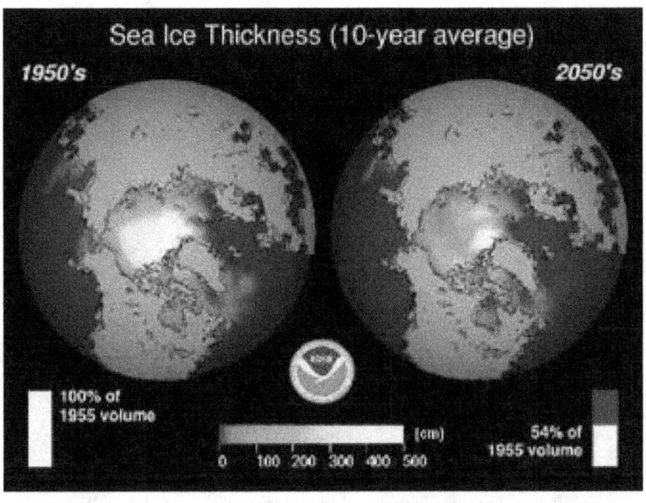

Fig. 5.1, Variation in sea ice thickness in Arctic-1950's - 2050's (Page 67)

Steps Towards A Green World

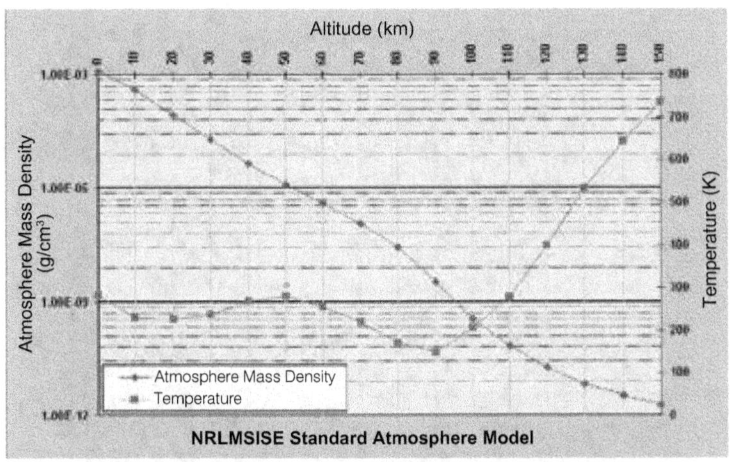

Fig. 18.2, Atmospheric mass density and temperatures against altitude
(Page 177)

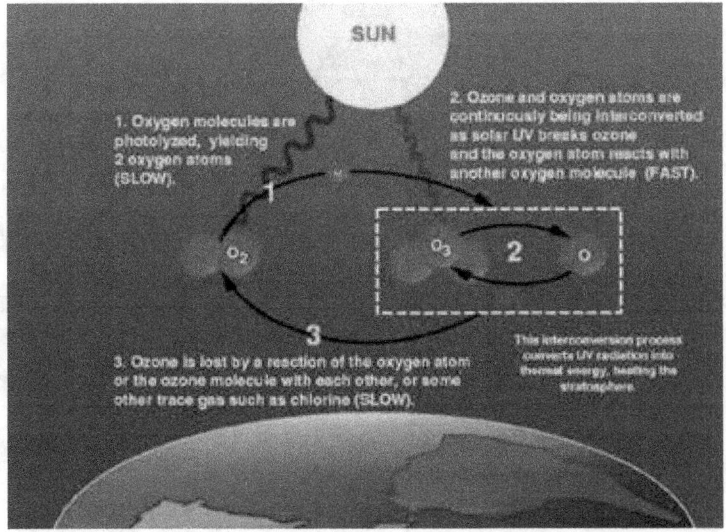

Fig. 20.1, Ozone-oxygen cycle in the ozone layer (Page 186)

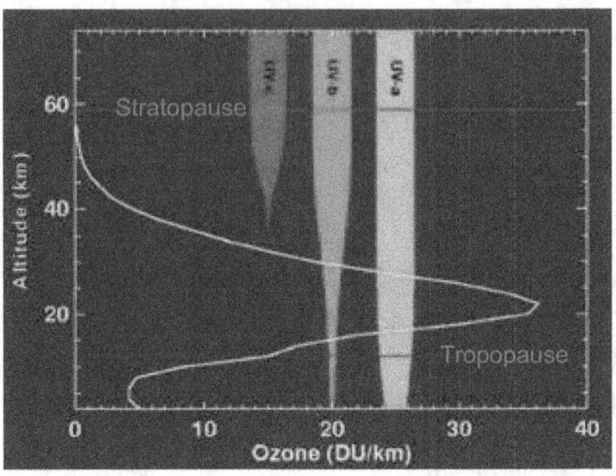

Fig. 20.2, Levels of ozone at various altitudes and blocking of ultraviolet radiation (Page 187)

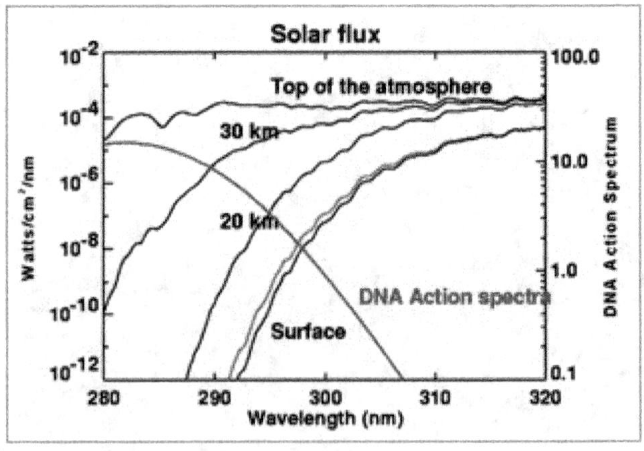

Fig. 20.3, DNA sensitivity to UV (Page 188)

Steps Towards A Green World

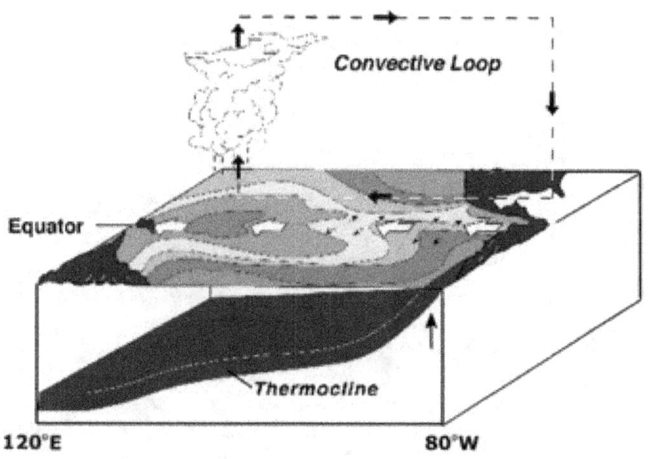

Fig. 22.1, Normal pacific pattern. Equatorial winds gather warm water pool toward west. Cold water up wells along South American coast. (NOAA / PMEL / TAO) (Page 200)

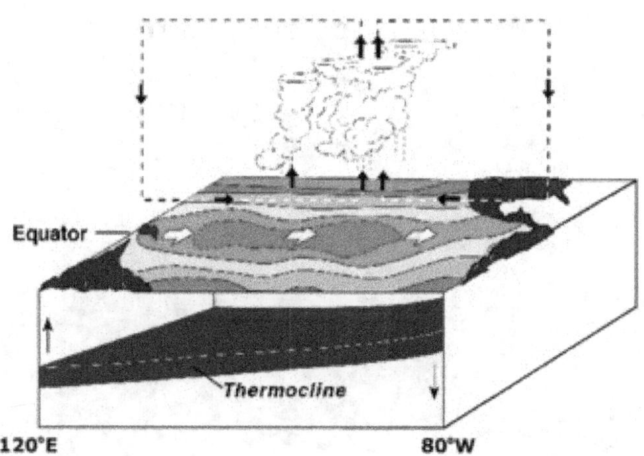

Fig. 22.2, El Niño conditions. Warm water pool approaches South American coast. Absence of cold upwelling increases warming (Page 201)

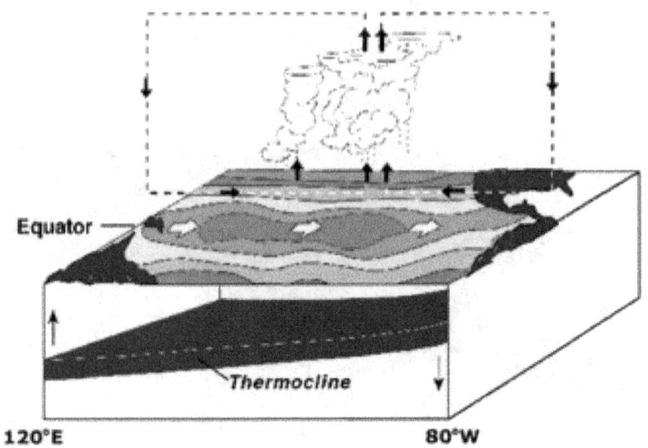

Fig.22.3, La Niña conditions. Warm water is further west than usual
(Page 201)

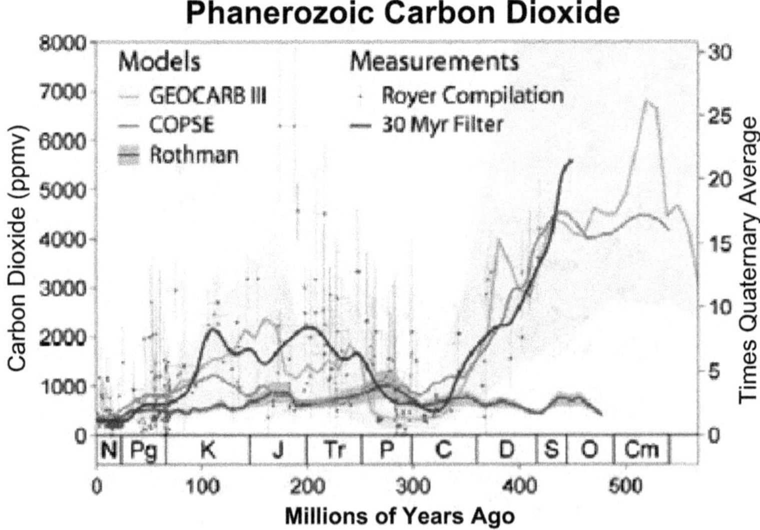

Fig.23.3, 500 million years of changes in carbon dioxide concentrations
(Page 209)

Steps Towards A Green World

ADDENDUM

DESI POWER FOUNDATION
Local Micro Grids
(An Innovative Solution)

Given below are excerpts, rephrased, from a presentation by Dr. H.N. Saran in March 2016.

Guiding Principal: Gandhiji's Vision of Indian Villages:

"I do not see why people should depend upon government ....If people depend upon themselves then the government is bound to move and this what I call democracy which is built up from below.

"You would not industrialise India?"

"I would indeed, in my sense of the term. The village communities should be revived".

Mission: Desi Power's promoters believe that the centralised sector alone will not be able to meet the development needs of Indian villages. Hence, it decided to promote self-reliant villages with decentralised, electricity-driven development processes integrated with job creation.

Objective: Desi Power Foundation's objective is to Pool and integrate resources of villagers, farmers and land owners for "sustainable rural development" to obtain Electricity, Water, Farming, Enterprises, Clean Cooking and Hygienic Safai Centres, for under-served villagers. Afore stated is shown in the schematic diagram ; "A Typical example of village level integration".

Activities and Projects

i) Micro Grids. Desi Power Foundation's Emppower Programmes, use **Renewable Energy Technologies (RET):** (i) Solar PV, (ii) Biogas, (iii) Hybrid (incl. solar +biogas or pumps). These projects help in building tiny grids of power plants in a large number of villages. A tiny grid generates about 1.2kWp serving 2-20 under-served persons. These supply, electricity for 335 days in a year. Tiny grids grow into Micro grids in 5-10 years. A Micro grid serves upto 0.8 km radius, comprising a large number of villages.

Tiny grids/ Micro grids provide domestic energy needs of lights, fans and small irrigation pumps.

Supply power for irrigation, agro-processing, energy needs for producing products for clean cooking, the power requirements of enterprises which help maintain and repair, locally, the installed plant and machines (electrical and mechanical), workshops, battery charging etc.

Green power is also supplied to 'Telecom Towers'.

ii) Establish Training Centres, DESI MANTRA, to provide training and skill development courses for villagers and project staff. The first centre was set up in 2006, at village Baharbari, in Araria District , Bihar. A training course was held as recently as February, 2016, at the training centre in Gairi village, Araria, Bihar.

iii) Kisanon Ki Bari (Farmers Garden). Their primary aims are: a) promote fair partnership between landowners and small farmers, b) increase farmers' income by crops of better productivity and higher value, e.g. mushrooms, fruits, c) arrange irrigation, bio-fertiliser and seeds, d) develop ponds for fish culture for better income of villagers.

iv) Establish plantations for Biomass; which can be processed for producing biogas for generating electricity, as well as clean fuel for domestic cooking.

 Steps Towards A Green World

v) Desi Gaon Safai Centres. Toilets and connected hygienic conditions.

vi) Desi Power Biomass Gasification Plants (for RET).

Three power plants have been commissioned, one in Madhya Pradesh (partially financed by the Swiss Foundation FRIEND), one in Orissa (partially financed by the Summit Foundation in the US) and one in Bihar (partially financed by the World Wildlife Fund - US). Currently, six IRPP (Karnataka, Andhra Pradesh, etc) plants are being built by DESI Power with financial support from the Government of Netherlands and the approval of the Government of India. Netherlands is the principal investor.

Future of Micro Grids: Governments have to put in place new policies, financing and incentives to encourage local or external private promoters to provide finance and help plan Micro grids. Thus, enable and encourage Panchayats and Gram Sabhas to pool and integrate resources of poor and under developed villages by increased farm production and creation of jobs locally. There will be virtually no CO_2 emissions and help eliminate fossil fuel based electricity.

Help meet the national goals of poverty reduction, social and ecological development co-currently with the global goal of reducing GHG emissions with lower direct costs.

Micro grids with decentralised renewable energy based power plants provide a real win-win solution at local, national and global levels.

Desi Power Foundation is a Trust established in 1996, by Dr. H.N. Saran. It is supported by Rockefeller Foundation and German, Swiss, Netherland and Indian NGOs.
Regd. Office: 32 Kempapura, Main Road, Bangalore, 560 024, Tel 08041328160
Head Office: Chitragupta Nagar, Ward No. 21, Araria-854311, Bihar, India.
Phone: 06453 222376, Mobile: 09334701127
Email: info@desipowerfoundation.org / arariaoffice@desipower.com/ aklavya@desipower.com
CEO: Deepak Das, deepak.das@desipowerfoundation.org

A typical example of village level integration

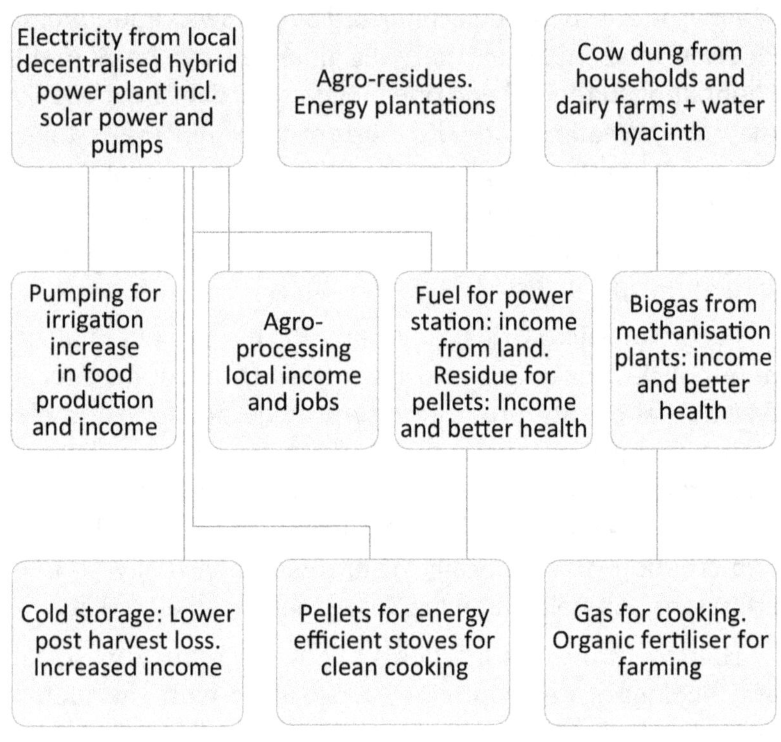

Addendum   241

www.ingramcontent.com/pod-product-compliance
Lightning Source LLC
Chambersburg PA
CBHW050326160726
48002CB00001B/203